AF333590

FEATURE INTERACTIONS IN TELECOMMUNICATIONS SYSTEMS, III

Feature Interactions in Telecommunications Systems, III

Edited by

K.E. Cheng and T. Ohta

Sponsored by

the IEEE Communication Society
the Institute of Electronics, Information and Communication Engineers of Japan
and the Advanced Telecommunications Research Institute International

1995

IOS
Press

Ohmsha

Amsterdam, Oxford, Tokyo, Washington, DC

ISBN 90 5199 238 6 (IOS Press)
ISBN 4 274 90065 7 C3055 (Ohmsha)
Library of Congress Catalogue Card Number 95-079020

Publisher
IOS Press
Van Diemenstraat 94
1013 CN Amsterdam
Netherlands

Distributor in the UK and Ireland
IOS Press/Lavis Marketing
73 Lime Walk
Headington
Oxford OX3 7AD
England

Distributor in the USA and Canada
IOS Press, Inc.
P.O. Box 10558
Burke, VA 22009-0558
USA

Distributor in Japan
Ohmsha, Ltd.
3-1 Kanda Nishiki - Cho
Chiyoda - Ku
Tokyo 101
Japan

LEGAL NOTICE

PRINTED IN THE NETHERLANDS

V

Contents

Committee

Workshop Co-Chairpersons
Tadashi Ohta (ATR, Japan)
Nancy Griffeth (Bellcore, USA)

Program Co-Chairpersons
Kong E. Cheng (Royal Melbourne Institute of Technology, Australia)
E. Jane Cameron (Bellcore, USA)

Members of Program Committee
Jan Bergstra (CWI and University of Amsterdam, The Netherlands)
Ralph Blumenthal (Bellcore, USA)
Rolv Braek (SINTEF DELAB, Norway)
Bernie Cohen (City University of London, UK)
Robert France (Florida Atlantic University, USA)
Haruo Hasegawa (OKI, Japan)
Dieter Hogrefe (University of Bern, Switzerland)
Richard Kemmerer (UCSB, USA)
Victor Lesser (University of Massachusetts, USA)
Yow-Jian Lin (Bellcore, USA)
Luigi Logrippo (University of Ottawa, Canada)
Jan van der Meer (Ericsson, The Netherlands)
Robert Milne (BNR, UK)
Leo Motus (Tallinn Technical University, Estonia)
Jacques Muller (CNET, France)
Jan-Olof Nordenstam (ELLEMTEL, Sweden)
Yoshihiro Niitsu (NTT, Japan)
Ben Potter (University of Hertfordshire, UK)
Henrikas Pranevicius (Kaunas University of Technology, Lithunia)
Martin Sadler (HP, UK)
Jean-Bernard Stefani (CNET, France)
Gregory Utas (BNR, USA)
Jyri Vain (Institute of Cybernetics, Estonia)
Hugo Velthuijsen (KPN Research, The Netherlands)
Yasushi Wakahara (KDD R&D Laboratories, Japan)
Ronald Wojcik (BellSouth, USA)
Pamela Zave (AT&T Bell Laboratories, USA)

Preface

This book is a compilation of the papers presented at the third Feature Interactions Workshop (FIW'95), held in the suburbs of Kyoto from 11 - 13 October, 1995, and is thus a continuation of an earlier volume, Feature Interactions in Telecommunications Systems, published in 1994.

Feature interactions happen when different features of a software system interfere with each other. The feature interaction problem has been most thoroughly documented in telecommunications systems, but it occurs in other systems as well. One interesting example of a feature interaction occurs in e-mail systems, between a mailing list program and a vacation program. The simplest way to implement a mailing list is to echo all mail to the entire list. But this implementation creates an unending cycle when a vacation program receives a message from the mailing list and sends back its automatic response. This actually occurred shortly after the feature interactions mailing list (fits-list@csi.uottawa.ca) was created.

Feature interactions appear to arise from the action of multiple collections of rules on an object. Often, these rules have been devised at different times or by different organizations, so that it is difficult to verify their combined effect. Because of the multiplicity, we might suspect that feature interactions are most likely in distributed systems, but they occur in single computer systems as well. For example, screen-savers on personal computers often will not run correctly in combination with certain graphics programs because each tries to control the display.

In an information society, heavily dependent on communications and distributed systems, feature interactions are likely to become an even more important problem than they are today. A particularly interesting issue, given the current work on agents, is whether feature interactions will be more likely in systems with many autonomous agents performing tasks.

The current demand for better and more convenient communications requires development of a variety of new services as quickly as possible. As the number of services becomes larger, however, feature interactions create incompatibilities between the various functions needed to implement them. In developing telecommunication systems, we now spend huge numbers of person-hours on software modifications and testing whenever a new function is added. Much of this time is spent detecting and eliminating problems arising from feature interaction. In the future, as ever more services are offered, feature interactions will become a major bottleneck in telecommunications-system software development.

Given this situation, the present workshop is being held as a positive opportunity to present and share opinions on the technical problems involved in feature interaction. In the first conference in this series, held in the United States in St. Petersburg, Florida, in December, 1992, a number of presentations were given, overviewing the understanding of feature interactions at the time and the applicability of computer technologies from a number of fields in solving the feature interaction problem. In the keynote address, Bob Martin (Bellcore) suggested the strategy of finding a tractable class of features, analogous to the LR(k) languages. The importance of developing precise definitions of features and feature interactions, and understanding the different classes of features and feature interactions, was also emphasized at the workshop.

Several papers selected from those presented at the workshop were published in the August 1993 special issue of IEEE Communications Magazine on feature interactions. In a simultaneous special issue, IEEE Computer published several papers on feature interactions describing potential contributions that computer scientists could make in various well-established computer science areas.

At the second conference, held in Amsterdam in the Netherlands, the focus was on concrete proposals for classifying feature interactions, and also for detecting and resolving them. Also at that second workshop, in a pre-workshop session held before the main workshop, free discussion was held on various topics such as the cooperative research by telecommunications-system researchers and computer scientists and the use of formal methods in addressing the feature interaction problem. A consensus of the workshop attendees was that the most difficult kind of feature to add to a system is one which is intended to modify a basic assumption of the original system, for example in the way that call forwarding modifies the assumption of a 1-1 relationship between a person and a telephone.

The present workshop will thus be the third, and is to be held at ATR, in the outskirts of Kyoto. This time the schedule is for three days. On the first, tutorials will be presented on the What are Feature Interactions, Formal Approaches to Feature Interactions, AIN and Feature Interactions, and Feature Interfaces and Feature Interactions. Time for free discussion based on these is also planned. In the technical sessions, the results of some of the latest research on feature interaction will be presented, and more detailed discussions will be expected.

FIW'95 Co-Chairpersons

Tadashi Ohta	and	Nancy Griffeth
ATR		Bellcore
Japan		USA

Introduction

New ways of dealing with the feature interaction problem continue to emerge, as reflected in the contributions to this workshop. With such a proliferation of approaches, we need to address an increasingly important issue of how we compare and assess these approaches in order to have some indications or objective measures of their effectiveness in dealing with the problem (or parts of). As a first step towards a comparative study in this area, we need to construct a reference framework in which existing and future approaches can be evaluated. Such a framework should include definitions, classifications and benchmarking of feature interactions, and associated assessment criteria. With such diverse viewpoints and interests, the development of an acceptable framework is a very challenging task. It could be taken as an ongoing process where the reference framework is continuously refined as we gain more insight into the problem. We hope this process will help us move towards a common and better understanding of the problem, and the eventual goal of eliminating the problem. We are sure, this and many other related issues will continue to receive interesting discussions in this workshop.

Contributions

In this workshop, the papers are divided into two categories: full presentation and short presentation. A short description of the papers is presented below.

Full Presentations

1. Automatic detection of feature interactions in temporal logic. This paper presents a restricted framework for the specification of features in temporal logic. The framework assumes that a system does not generate additional internal or outgoing events. Feature interaction is defined as the occurrence of an event that triggers transitions in two features with different effects. The result is a formal approach that supports automatic detection of selected feature interactions.

2. A formal framework for feature interaction with emphasis on testing. This paper presents a three-level formal framework for feature interactions - requirement, specification and implementation levels. The author identifies an interaction as originating from one of the three levels and may be passed on from one level to the next. The framework uses testing as the main mechanism for detecting interactions.

3. Issues of non-monotonicity in feature-interaction detection. This paper highlights the limitation of most existing formal approaches to dealing with feature interactions, namely the qualification and frame problems. The author shows that extending a telecommunications system with new features is a non-monotonic operation and asserts that non-monotonic techniques are the way forward to dealing with feature interactions.

4. Runtime resolution of feature interactions in architectures with separated call and feature control. This paper describes a three-part technique to resolve feature interactions based on roles and finite state machines. The first involves comparing compatibility of feature roles. The second part introduces a feature state machine that ensures an event will only be assigned to a feature instance if it leads to a desired state transition of the feature. The last part defines rules for event distribution. The result is a runtime resolution technique for detecting interactions.

5. Minimizing feature interactions: an architecture and processing model approach. This paper describes an approach to minimise feature interactions based on an architecture that uses software agents. The architecture eliminates selected interactions by removing common assumptions about roles and technology. Associated with the architecture, a processing model is proposed to ensure independent management of feature interactions.

6. An interaction-avoiding call processing model. This paper describes a call processing model that avoids feature interactions. The model uses roles and sub-roles to distinguish network users from the equipment they use. Interactions are avoided by specifying call control requirements of all participating parties at call setup.

7. A dynamic resolution method for feature interactions and its evaluation. This paper presents an interaction resolution method using a protocol synthesis technique. Features are defined as a set of predicates in the service specification. Interactions are resolved at the protocol specification level i.e. through the process of synthesis from the service specification to the protocol specification. The paper concludes by applying the approach to a set of telephone services.

8. Automatic reconfiguration for runtime feature-interaction resolution in an object-oriented environment. This paper uses an object-oriented approach to resolve feature interactions. The author identifies two factors associated with the interaction problem. The first is the lack of a complete and formal definition of the boundary between a system and its environment. The second factor is a weak description of configuration states - states of resources in a system. The result is a resolution approach that deals with interactions across system-environment boundary at a given re-configured state.

9. A new proposal for feature interaction detection and elimination. This paper identifies the limitations of an earlier work that uses a finite state machine approach to detect interactions. The deficiencies: timing verification, verification of retrials (such as those caused by errors in user inputs), and verification of synonyms (semantically identical events and states) are overcome in a new method expressed in Petri nets.

10. Representing and verifying intentions in telephony features using abstract data types. This paper uses an abstract data type (ADT) approach to specify the intentions of features. Violations are specified as negations (indirectly) of intentions using ADT operations i.e. by reversing boolean expressions of the original intentions. Verification of intentions is based on detecting the existence of violations. The approach is represented (partially) in three languages: SDL, LOTOS and Prolog.

11. Feature interaction detection and resolution in the Delphi framework. The Delphi language is intended to be used at the beginning of the service creation process to produce an implementation independent service specification. The static part of a Delphi language consists of entity types of a system and their relationship; while the dynamic part is a set of rules expressed as state transition functions. Feature interactions are detected by identifying unsound rules i.e. those that lead to inconsistent conclusions. Interactions are resolved by ensuring all rules are mutually exclusive i.e. no inconsistent conclusions.

Short Presentations

12. Service validation and testing. This paper describes a practical approach to validate and test Intelligent Network (IN) services. The approach uses two models: network and service plane models. The network model supports several nodes with an in-built interactions detection ability in its architecture. The service plane model is an abstract view of the telephone service with mechanisms for creation and addition of services. The approach is modelled using SDL.

13. A platform for modelling feature interaction detection and resolution techniques. This paper describes a platform that is used to support the development of experimental call models such the IN call model. The platform could be used as a testbed to investigate the feature interaction problem.

14. Towards a more efficient feature interaction analysis - a statistical approach. This paper outlines an approach to guide feature interaction analysis based on a statistical approach. The method advocates the use of system usage distribution as a more efficient way of verification.

15. Resolving service interactions by service components. This paper presents an interaction resolution technique using a feature interaction manager that acts as an intermediary between service logics and a basic call model. The approach is based on the ITU-T IN conceptual model and is specified using SDL.

Towards a comparative framework

In general, the process of dealing with the feature interaction problem can be summarised into three aspects: viewpoint analysis, problem definition and development of possible solutions.

Viewpoint analysis examines the feature interaction problem from a particular perspective of interest. For instance a user's view and interests of a telephone network is different from those of a network operator. A number of possible viewpoints are presented in [1].   Based on a selected viewpoint, the problem is defined in terms of how it could be classified and represented. Classification of feature interactions during problem definition could vary from an ad hoc approach (such as a random collection of feature interaction cases) to the development of a complete taxonomy. Sometimes the problem is represented using formal techniques. Examples of classifications and formal representations can be found in [1,2].

Once the problem is defined, we proceed with the development of possible solutions based on our particular view of the problem. The solutions usually involve using informal or formal techniques to avoid, detect and/or resolve feature interactions [2]. Avoidance examines ways of preventing feature interactions. Detection looks at techniques for locating feature interactions. Resolution examines methods for eliminating feature interactions once they are detected.

As an exercise, we have grouped the contributions in this workshop based on the three aspects outlined above with the viewpoints identified in [1]: software life cycle view (SLC), network configuration view (NC), causal view, layered view and organisational/operational view (OGOP). We are interested in looking at the types of contributions in this workshop, and obviously a more comprehensive framework is needed to further develop this area of work. The contributions are summarised in Table 1.

Table 1. Summary of Contributions

No	Viewpoint		Problem Definition		Solution	
	View	Context	Classification	Representation	Approach	Representation
1	SLC/causal	- requirement level	- pre-conditions - post-conditions	- temporal logic	- detection	- temporal logic
2	SLC/causal/ layered	- requirement level - specification level - implementation level	interactions at and between pairs of requirement, specification and implementation levels	- req. (temporal logic) - spec. (state machine or process algebra) - implementation relations	- detection	- testing language (TTCN)
3	SLC/causal	- non-monotonic behaviour			- detection	- non-monotonic techniques
4	NC/layered	- architecture and platform	- functional/roles (semantic) - event (syntactic)	- roles - state machine	- resolution	- roles - state machine
5	NC/layered	- architecture and platform		- agents (user, service, connection & resource)	- detection - resolution	- agents (user, service, connection & resource)
6	NC/layered	- architecture and platform	issue-oriented configuration /causal taxonomies [1]	- Z - standard ML	- avoidance	- Z - standard ML
7	layered	- protocol synthesis - com. architecture	- service specification - protocol specification	- state machine	- resolution	- graph
8	NC/layered	- system-env. boundary - states of system resources	- system-env. boundary - states of system resources	- objects and objects interactions	- resolution	- object-oriented environment
9	SLC	- specification level		- state machine - petri nets	- detection - resolution	- state machine - petri nets
10	SLC/causal	- specification level	- feature intentions - feature violations	- ADT (LOTOS, SDL)	- detection	- ADT (LOTOS, SDL) - Prolog
11	SLC/causal	- specification level		- entity types and relations - state transition functions	- detection - resolution	- entity types and relations - state transition functions
12	NC/layered	- service plane model - Centrex network model		- SDL	- detection - resolution	- SDL
13	layered	- architecture and platform				
14	SLC	- requirement level	efficiency of analysis	- statistic (service usage)		
15	layered	- architecture and platform		- SDL	- resolution	- SDL

Conclusions

This workshop continues to reflect on a growing number of approaches that deal with the feature interaction problem. It is encouraging to see ongoing interests in attempting to solve this important issue from many different viewpoints, and in working towards a better understanding of the different approaches and the problem. In particular, we need to improve the way we define and classify feature interactions, and to put more effort into evaluating, comparing and/or integrating existing (and future) approaches. From an application viewpoint, we need to provide software architectures that enable features to be more easily integrated into telephone networks. A well-defined software architecture will facilitate the process of applying existing research results to the deployment of services and features.

References

[1]   E.J. Cameron and H. Velthuijsen. Feature interactions in telecommunications systems. IEEE Communications Magazine, Vol. 31 No. 8, August 1993

[2]   W. Bouma and H. Velthuijsen (Eds.). Feature interactions in telecommunications systems, IOS press, 1994.

FIW'95 Program Co-Chairpersons

Kong E. Cheng   and   E. Jane Cameron
Royal Melbourne Institute of Technology   Bellcore
Australia   USA

Acknowledgments

The editors would like to thank the authors for their contributions and members of the program committee for reviewing the papers. In particular, we would like to mention the effort of the workshop co-chair Dr. Nancy Griffeth, program co-chair Dr. Jane Cameron and Dr. Hugo Velthuijsen of KPN Research. We would also like to acknowledge the sponsorship of ATR, especially the support of the Chairman of ATR, Dr. Kohei Habara and the President of ATR, Dr. Yoshinobu Terashima.

Editors

Kong E. Cheng   and   Tadashi Ohta
Royal Melbourne Institute of Technology   ATR
Australia   Japan
Tel: +61 39 660 3266   Tel: +81 774 95 1230
Fax: +61 39 662 1617   Fax: +81 774 95 1208
Email: kec@cs.rmit.edu.au   Email: ohta@atr-sw.atr.co.jp

Feature Interactions in Telecommunications Systems III
K.E. Cheng and T. Ohta (Eds.)
IOS Press, 1995

Automatic Detection of Feature Interactions in Temporal Logic

Johan Blom, Roland Bol, and Lars Kempe
Uppsala University
Dept. of Computer Systems
P.O. Box 325, S-751 05 Uppsala, Sweden
{johan,rolandb,larsk}@docs.uu.se

1 Introduction

We present a uniform framework for the specification of telephone systems in linear-time temporal logic [MP92, Lam91]. This logic can be used to specify high-level properties (cf. Section 5), but also for a more implementation oriented style of specifications, similar to [But93, GK94]. It is on the later style we focus here. We assume that a system has a set of *variables*, which determine its *state*. The system reacts to incoming *events* by changing its state. It does not itself generate further (internal or outgoing) events. (We are aware of the fact that this is a serious restriction; we return to it in Section 6.) A specification in this style defines a set of *transitions*, triggered by events. *Interaction* (sometimes called interference) between features occurs when an event triggers transitions in both features that specify different effects (again, similar to [GK94]).

Our motivation for considering this style is twofold. First, we can exploit the restricted framework to detect feature interactions automatically. This idea is the core of the paper. Second, this level of description is probably most realistic. Certainly, when a new feature is proposed, it will exist at first as a collection of informal requirements. In accordance with common practice in the telecommunications industry, these requirements turn out to be easiest to formalise by the kind of state machine we propose.

Because our specifications are written in temporal logic, we can characterise states in which an event causes interaction by a state formula. The existence of such states does not automatically imply that there is feature interaction: these states might not be reachable. At this point, we invoke the help of the tool described in [JK95], which, given a transition system and a state formula in a particular form, attempts to decide whether a state that satisfies the formula is reachable. We say 'attempts' because this problem is undecidable in general. The usefulness of this tool is subject to further study.

Throughout the paper, we use the specification of a simple telephone system as an example. In Section 2.2 we specify the basic system POTS (Plain Old Telephone System), which we extend in Section 4 by (a partial specification of) two features. Then we apply our methodology to detect several interactions. Resolving the interactions, once detected, is beyond the scope of this paper; we give some hints on this topic in the final section. In the same section, we discuss how our current view of the system should be refined in order to arrive at the distributed telecommunication networks that exist in practice.

2   The Framework

In this section, we introduce our specification framework. As a running example, we use a specification of POTS. A discussion regarding the expressiveness of the framework concludes this section.

2.1   Specifications

Throughout the paper, we will employ the following general specification style. A system (or additional feature) is specified by a formula

$$Initial \wedge \Box\, \forall\, React,$$

where *Initial* is a state formula specifying the initial values of the state variables of the module and *React* specifies the reactive behaviour by a conjunction of formulas (*reactive rules*) of the form

$$event \Rightarrow allowed$$
$$\wedge\ condition_1 \overset{id_{s1}}{\Rightarrow} effects_1$$
$$\wedge\ condition_2 \overset{id_{s2}}{\Rightarrow} effects_2$$
$$\vdots$$

In this formula, *event* is the name of the incoming event and its parameters, if any. *allowed* is a state formula: it characterises exactly those states in which this event can happen. For example, in POTS we have a state variable *active* that keeps track of the position of the receiver. The fact that it is impossible to lift the receiver if it is already lifted, is represented by an allowedness condition. The *condition*s are also state formulas. They allow a case distinction on the current state to determine the effects of the event. They need *not* be exclusive. Finally, the *effects* are conjunctions of equations of the form $variable^+ = expression$, giving the value of a state variable in the next state in terms of the values of variables in the current state.

Our state variables represent relations between *nodes*. In POTS, these nodes are (potential) subscribers of the system: $A \in active$, usually written as $active(A)$, denotes that subscriber A has lifted his receiver. Reactive rules do not normally mention individual subscribers like A: they apply to all subscribers. Thus we shall only encounter variables for nodes, usually denoted by id, id'. Events have normally at least one node as a parameter: the node where the event occurs. Of course, they can have more parameters. The universal quantification in $\Box\, \forall\, React$ quantifies over these parameters.

It can happen that the condition contains an existentially quantified variable id that we want to refer to in the effect. For example, a specification in natural language might say: 'If someone is calling you, then answer him.' Sentences like this one are a well known problem for formalising natural language in logic. The word 'someone' captures not only an existentially quantified variable in the if-part, but also declares an 'entity' to which 'him' in the then-part can refer. We solve this problem by allowing the condition to declare the entities *ids*, to which the effect part can refer. Formally,

$$condition \stackrel{ids}{\Rightarrow} effects \;\widehat{=}\; (\exists\, ids : condition) \Rightarrow (\exists\, ids : condition \wedge effects).$$

Note that when the number of *ids* is zero, both the notation and its semantics reduce to ordinary implication.

2.2   Example: POTS

In POTS, we have the unary state variables *active* and *busy*, and binary state variables *trying* and *connected*. Initially, all these relations are empty:

$$POTSInit \;\widehat{=}\; active = busy = trying = connected = \varnothing.$$

For readability, we abbreviate the following state formulas:

$$IsBusy(id) \;\widehat{=}\; active(id) \vee \exists\, id' : trying(id', id).$$

$$\begin{aligned}
CanDial(id) \;\widehat{=}\; &active(id) \wedge \neg\, busy(id)\\
&\wedge \nexists id' : trying(id, id')\\
&\wedge \nexists id' : connected(id, id')\\
&\wedge \nexists id' : connected(id', id).
\end{aligned}$$

POTS recognises three events: *Offhook*(*id*) means that *id* lifts the receiver (either to be able to dial or to answer a call), *Onhook*(*id*) means that *id* hangs up, and *Dialling*(*id*, *id'*) means that *id* dials *id'* (so that either *id'* is called, or *id* gets a busy tone). These intuitions are reflected by the following reactive rules (in which the layout replaces some brackets; the notations ◁/▷ stem from Z [Spi89] and denote domain/range restriction).

$POTSReact \; \widehat{=}$

$Offhook(id) \Rightarrow \neg\; active(id)$

$$\begin{aligned}
&\wedge && active^+ &&= active \cup \{id\} \\
&\wedge && busy^+ &&= busy \\
&\wedge \; \nexists id' : trying(id', id) \Rightarrow\; trying^+ &&= trying \;\wedge \\
&&& connected^+ &&= connected \\
&\wedge \; trying(id', id) \quad \overset{id'}{\Rightarrow}\; trying^+ &&= trying \setminus \{\langle id', id\rangle\} \;\wedge \\
&&& connected^+ &&= connected \cup \{\langle id', id\rangle\}
\end{aligned}$$

$\wedge \; Onhook(id) \Rightarrow active(id)$

$$\begin{aligned}
&\wedge && active^+ &&= active \setminus \{id\} \;\wedge \\
&&& busy^+ &&= busy \setminus \{id\} \;\wedge \\
&&& trying^+ &&= \{id\} \vartriangleleft trying \;\wedge \\
&&& connected^+ &&= \{id\} \vartriangleleft connected \vartriangleright \{id\}
\end{aligned}$$

$\wedge \; Dialling(id, id') \Rightarrow CanDial(id)$

$$\begin{aligned}
&\wedge && active^+ &&= active \\
&\wedge && connected^+ &&= connected \\
&\wedge \; IsBusy(id') &&\Rightarrow busy^+ &&= busy \cup \{id\} \;\wedge \\
&&& trying^+ &&= trying \\
&\wedge \; \neg\; IsBusy(id') &&\Rightarrow trying^+ &&= trying \cup \{\langle id, id'\rangle\} \;\wedge \\
&&& busy^+ &&= busy
\end{aligned}$$

2.3 Datastructure

In Section 3.2, we describe a tool that can carry out reachability analysis for just the kind of reactive rules defined in Section 2.1, provided that they are phrased in a particular datastructure, which we define now. POTS satisfies this requirement.

- The state variables are *unary* and *binary* relations between a fixed set of nodes.

- The parameters of an event are nodes. No node constants occur in reactive rules, only node variables (such as id and id').

- The state formulas *allowed* and *condition* are conjunctions and disjunctions of

    - positive and negative *simple literals*, i.e. relations about and between nodes for example, $active(id)$ and $\neg connected(id, id')$.

    - positive and negative *existential literals*, i.e. existentially quantified positive simple literals for example $\exists id' : trying(id, id')$.

- The *effect* formulas can

    - add particular (pairs of) nodes to a relation ($active^+ = active \cup \{id\}$),

    - remove particular (pairs of) nodes from a relation ($trying^+ = trying \setminus \{\langle id', id\rangle\}$),

    - remove from a relation all pairs of nodes that have a particular first and/or second component (denoted by e.g. $trying^+ = \{id\} \vartriangleleft trying$, as in Z).

    - leave a relation unchanged.

In other words, an *effect* formula can be identified with a conjunction ψ of positive and negative simple literals, and negative existential literals. Namely, it specifies that the next state is the same as the current state, *except* that now ψ holds. For example, the effect $trying^+ = \{id\} \triangleleft trying$ is represented by the negative existential literal $\not\exists id' : trying(id, id')$. Positive existential literals are not allowed in an *effect* formula, because they fail to determine a unique effect.

2.4 Unchanged variables

In POTS, most events can change most state variables. In a large, modular system, we expect the opposite: events typically change only a few of many variables. As there is no *frame axiom* ('nothing changes unless it has to') in our logic, we must specify explicitly that an event cannot change a variable. For one feature, this might be cumbersome, but it is always possible. However, we want to add new features, with new events and new variables, in a modular way. That is, without adding to each new reactive rule that its events does not change the old variables (and vice versa).

Therefore we add a third component to our framework, stating that particular state variables can *only* be changed by particular events. We call such statements *frame formulas*. In the following, $\text{Changed}(x_1, \ldots, x_n)$ abbreviates $x_1^+ \neq x_1 \vee \ldots \vee x_n^+ \neq x_n$. Frame formulas have the form

$$\text{Changed}(x_1, \ldots, x_n) \Rightarrow \exists\, event_1 \vee \ldots \vee \exists\, event_k.$$

This formula states that *only* the events $event_1$ and ... and $event_k$ can change the variables $x_1, \ldots, x_n$. The existential quantification refers to parameters of the events. Thus a complete specification is now of the form

$$System \;\hat{=}\; Initial \wedge \Box\, \forall\, React \wedge \Box\, Frame,$$

where *Frame* is a conjunction of frame formulas. An important observation is that *composition of specifications can be done by conjunction*: by some simple distributivity laws, one can see that the conjunction of specifications is again a specification.

For example, the frame formulas for POTS are

$$POTSFrame \;\hat{=}$$

$$
\begin{array}{ll}
\text{Changed}(active, connected) & \Rightarrow \exists\, id : Offhook(id) \qquad \vee\, \exists\, id : Onhook(id) \\
\wedge\, \text{Changed}(trying) & \Rightarrow \exists\, id : Offhook(id) \qquad \vee\, \exists\, id : Onhook(id)\, \vee \\
& \quad\; \exists\, id, id' : Dialling(id, id') \\
\wedge\, \text{Changed}(busy) & \Rightarrow \exists\, id, id' : Dialling(id, id') \vee \exists\, id : Onhook(id).
\end{array}
$$

Note that they refer only to events and variables in POTS, and can therefore be stated without knowing which additional features are going to be added. With these frame formulas, we could omit the effects $busy^+ = busy$ in *Offhook* and the first two effects in *Dialling*.

2.5 Discussion

The framework we defined raises a number of questions.

- About the framework itself:

- What kind of systems can we specify in this framework?
- What are desirable properties of specifications?

These questions are discussed below.

- About detecting interaction:

  - Which forms of interaction can be distinguished?
  - How are they characterised logically?
  - Why is this framework suitable for automatic detection, and how is it done?

These questions are central to the rest of this paper.

- About avoiding and resolving interaction:

  - In which cases is it inevitable that a new feature interacts with the existing system?
  - How can we resolve such interactions systematically?
  - How can we resolve other ('real') interactions?

These questions are slightly outside the scope of this paper. Nevertheless, we address them briefly in the concluding section, in order to validate the applicability of our framework.

The restrictions on the datastructure in our framework seem more severe than they really are. In fact, one can easily simulate a Turing machine by a finite set of *condition* $\Rightarrow$ *effect*-rules. The limitations of our framework are not on the level of computational power, but on the level of expressiveness.

When an event is allowed, but the conditional effects do not specify the new value of a variable after the event (and the event can change the variable according to the frame formulas), then the new value of the variable can be anything. This is a result of the absence of the frame axiom. This form of nondeterminism seems not useful, even undesirable, in specifications. A potentially more useful form of nondeterminism can be observed in the last conditional effect of *Offhook*: one of the id's trying to call id is selected for a connection. Of course, in the current specification there can be only one such id', but in general there could be more (for example by dropping $\exists\, id' : trying(id', id)$ from *IsBusy*).

The effects of an event are limited to an immediate state change, and cannot generate follow-up events. In particular, we cannot specify liveness properties, which require looking further into the future than one state. Finally, as will be obvious from the example, we do not handle concurrently incoming events. We return to these limitation in the last section.

It is possible to write 'contradictory' specifications, for example by writing two rules assigning different allowedness conditions to the same event (the result is that the real allowedness condition becomes their conjunction), or by specifying inconsistent effects for an event (the result is that the event is not allowed). We assume in the rest of this paper that single specifications are 'well-formed', so that when dealing with interaction between specifications, we need not worry about such problems within a specification. Attempting to define 'well-formed' more precisely here turns out to be almost the same as defining 'interaction', which we do in the next section: interaction occurs if the composition (conjunction) of two well-formed specifications is not well-formed (note the similarity with the properties discussed in Section 5.)

3 Interaction Detection

In this section, we assume specifications S_1 and S_2, with

$$S_1 \triangleq Initials_1 \wedge \Box \, \forall \, React_1 \wedge \Box \, Frame_1,$$

and similar for S_2. Furthermore, we assume that the parameters of events are *consistently named*, that is, if $event(p_1, \ldots, p_n)$ occurs in S_1, then *event* occurs in S_2 only with the exact same list of parameters. These assumptions are just for convenience. They mean that event names are not overloaded and that we assume that the required α-conversions have been carried out. As the number of parameters of an event is usually unimportant, we write $event(p)$ for $event(p_1, \ldots, p_n)$. Similarly, we represent the state variables by just x.

3.1 Forms of Interaction

Several ways of classifying feature interactions can be found in the literature:

- By location: the number of users and components that are involved (Section 2 of [CGN$^+$93]),

- By cause: timing, assumptions about the database, limited resources, and so on (Section 3 of [CGN$^+$93], Section 3.3 of [BA94]),

- By design stage: the aims of the features can be incompatible, or their implementation on the level discussed in this paper, or on the level of signals and resources (Section 2 of [OH94]),

- By formalism: a specification framework can 'inform' us in different ways about feature interactions (Section 3 of [OH94]).

In this section, we present a classification based on our formalism.

Each specification consists of three components, so we might expect six different kinds of interaction. In fact, only two are interesting. We consider a possible inconsistency of $Initial_1 \wedge Initial_2$ as uninteresting. Furthermore, the initial formulas cannot contradict the other parts of the specifications. Finally, two frame formulas cannot be contradictory. That leaves two possible interactions: between two reactive rules and between a reactive rule and a frame formula. We shall first investigate interactions between two reactive rules, and return to the frame formulas at the end of the section.

Say we have an event with parameters $event(p)$ present in both specifications, and S_1 contains the reactive rule

$$event(p) \Rightarrow allowed_1(x, p)$$
$$\wedge \, c_1(x, p, ids_1) \overset{ids_1}{\Rightarrow} e_1(x, p, ids_1, x^+)$$

and a similar rule in S_2. Thus we separate the allowedness condition from the conditional effects.

The first possibility of interaction occurs if the two specifications disagree about the allowedness of an event. Formally, this can be expressed as

$$\exists \, p : \neg \, (allowed_1(x, p) \Leftrightarrow allowed_2(x, p)).$$

We say 'possibility', because even if there is a state x in which the above formula is true, this state might not be reachable from the initial state.

A second possibility is that the event is allowed, but that some of the conditional effects contradict. Formally,

$$\exists\, p, ids_1, ids_2 : allowed_1(x, p) \wedge c_1(x, p, ids_1) \wedge c_2(x, p, ids_2)$$
$$\wedge\ \not\exists x^+ : e_1(x, p, ids_1, x^+) \wedge e_2(x, p, ids_2, x^+).$$

Again, such states x might exist, but not be reachable. In the next section, we describe a tool that tries to decide such reachability questions (it can only try: these questions are in general undecidable).

For the purpose of detecting interactions, one can show that there is no logical difference between the parameters p and the existentially quantified variables *ids* inside the rules. Therefore we can assume that any *ids* in a rule is part of the p for the rule.

3.2 A Tool for Reachability Analysis

We give an overview of the reachability procedure described in [JK95]. A prototype implementation of the procedure has been written in Prolog. It takes as input a transition system $TS = (I, T)$, where I is the initial state and T is a set of transitions, and a state formula (*pattern*) ρ_0 that characterises 'bad' states (for example: states in which feature interaction occurs). Then it weakens ρ_0 to ρ_1, which characterises all states from which a bad state is reachable in at most one step. It continues by computing $\rho_2, \ldots$, until the initial state satisfies some ρ_n (a bad state is reachable), or the sequence of ρ_is reaches a fixed point (ρ_n and ρ_{n+1} are logically equivalent). If neither happens, then the procedure does not terminate. This approach was inspired by [AJ93].

How ρ_{i+1} is computed from ρ_i and TS, how it is determined if I satisfies ρ_i and how it is detected that a fixed point is reached can be found in [JK95]. Below we describe the restrictions imposed by the tool. The datastructure of our framework, as described in Section 2.3, is adapted to these restrictions in three ways.

- The initial state I is uniquely determined by the *Initial* formula (it is easy to get around this requirement, if necessary).

- The formulas *allowed*, *condition*, and *effect* define indeed transitions.

- The formulas characterising interaction are of the kind that the tool can handle.

We use the terminology of Section 2.3. The states of the transition system are conjunctions of positive simple literals. Intuitively, these literals are *exactly* the ones that hold: others do not hold. A transition is a pair (φ, ψ), where φ is the precondition and ψ is the postcondition. The precondition is a conjunction of positive and negative simple literals and positive and negative existential literals. The postcondition is a conjunction of positive and negative simple literals and negative existential literals. Their interpretation is the one outlined in Section 2.3.

From a specification in our framework we can derive a set of transitions as follows.

Each reactive rule[1]

$$event \Rightarrow allowed$$
$$\wedge \bigwedge_{i=1\ldots n} c_i \Rightarrow e_i$$

in the specification yields a set of transitions. Each of these transitions is characterised by the set $S \subseteq \{1, \ldots, n\}$ of conditional effects $c_i \Rightarrow e_i$ that 'fire', i.e., have their condition c_i satisfied. The condition *allowed* must of course always be satisfied, regardless from S. Thus, under the precondition

$$\varphi_S \triangleq allowed \wedge \bigwedge_{i \in S} c_i \wedge \bigwedge_{i \notin S} \neg\, c_i$$

we can obtain the effects (postcondition)

$$\psi_S \triangleq \bigwedge_{i \in S} e_i.$$

By rewriting the above precondition φ_S into a disjunctive normal form $\varphi_S^1 \vee \ldots \vee \varphi_S^n$, we obtain transitions $(\varphi_S^1, \psi_S), \ldots, (\varphi_S^n, \psi_S)$. Doing this for each reactive rule, and for each subset of its conditional effects, we obtain the complete transition system.

The above construction is necessary, because we have allowed that the conditions in our framework are not mutually exclusive and can contain disjunctions, whereas this is not allowed by the tool. This serves to make the specification more readable. For example, the effect of *Offhook* on *active* does not depend on the case distinction that affects *trying* and *connected*. The worst-case complexity of the construction is obviously high, but in general the resulting transition system has the same size as the specification. Namely, a reactive rule usually consists of some unconditional effects, and one or two case analyses for the other effects. For most selections S, the precondition φ_S will therefore be inconsistent (and thus not result in any transitions). The transition system resulting from the specification of POTS in Section 2.2 is given in Table 1 as an illustration.

One could even consider to forget about the logical framework altogether, and specify the transition system directly, as in [ITO92]. Then the question is of course how to arrive at a characterisation of interactions corresponding to the analysis in the previous section. In particular for nondeterministic specifications, analysis based directly on transition systems is hard. Moreover, we could not simply refer to the semantics of temporal logic as a semantics for our specifications. Finally, we would lose the possibility of later specifying high-level properties in the same language as our specifications.

It remains to define the kind of formulas that the tool can handle. These formulas, called *patterns*, are existentially quantified precondition formulas, with the additional restriction that the quantified variables are all disjoint. For example, we verified for the above transition system for POTS the invariance of the property that two subscribers can only be connected if they are both active; formally:

$$\forall\, id, id' : connected(id, id') \Rightarrow active(id) \wedge active(id').$$

This was done by taking the negation of the above formula, and rewriting the formula in disjunctive normal form. The result is the disjunction of three patterns:

$$\exists\, id, id' : id \neq id' \wedge connected(id, id') \wedge \neg\, active(id),$$
$$\exists\, id, id' : id \neq id' \wedge connected(id, id') \wedge \neg\, active(id'), \text{ and}$$
$$\exists\, id : connected(id, id) \wedge \neg\, active(id),$$

---

[1]As pointed out in the previous section, we do not need to take into account separately the conditional effects $condition \overset{ids}{\Rightarrow} effects$, because we can consider *ids* as additional parameters of the event. For example, *id* and *id'* have the same status in the second transition of Table 1, the trigger event is not relevant for the tool.

event	φ	ψ
$\mathit{OffHook}(id)$	$\neg\ active(id)$ $\nexists id' : trying(id', id)$	$active(id)$
	$\neg\ active(id)$ $trying(id', id)$	$active(id)$ $\neg\ trying(id', id)$ $connected(id', id)$
$\mathit{OnHook}(id)$	$active(id)$	$\neg\ active(id)$ $\neg\ busy(id)$ $\nexists id' : trying(id, id')$ $\nexists id' : connected(id, id')$ $\nexists id' : connected(id', id)$
$\mathit{Dialling}(id, id')$	$CanDial(id)$ $active(id')$	$busy(id)$
	$CanDial(id)$ $\exists id'' : trying(id'', id')$	$busy(id)$
	$CanDial(id)$ $\neg\ active(id')$ $\nexists id'' : trying(id'', id')$	$trying(id, id')$

Table 1: The transition system for POTS.

which were shown not to be reachable. The addition of the third pattern shows how we can always get around the disjointness condition on patterns by adding an extra pattern. More properties of POTS were verified in the same way.

In the next section we show that disjunctions of patterns can express the two kinds of possible interactions described in Section 3.1.

3.3   Reachability of Interactions

As we argued in Section 3.1, in order to detect that two specifications disagree on the allowedness condition of an event, we must compute if there are states x that satisfy

$$\exists p : \neg\ (allowed_1(x, p) \Leftrightarrow allowed_2(x, p).$$

By writing this formula in disjunctive normal form, we obtain a set of patterns, which can be checked for reachability by the tool.

We also have an interaction when two conditional effects of an event conflict:

$$\exists p : allowed_1(x, p) \wedge c_1(x, p) \wedge c_2(x, p) \wedge \nexists x^+ : e_1(x, p, x^+) \wedge e_2(x, p, x^+),$$

or more precisely, when they conflict for any variable v in the tuple x of state variables:

$$\exists p : allowed_1(x, p) \wedge c_1(x, p) \wedge c_2(x, p) \wedge \bigvee_{v \text{ in } x} \nexists v^+ : e_1(v, p, v^+) \wedge e_2(v, p, v^+).$$

This formula is not a (disjunction of) patterns, so the tool cannot be used directly on it. The problem is caused by the subformula $\nexists v^+ : e_1(v, p, v^+) \wedge e_2(v, p, v^+)$, which is essentially universally quantified.

However, the formulas $e_{1,2}(v, p, v^+)$ must be assignments of the form $v^+ = e(v, p)$. Moreover, e is not just any expression, it can be regarded as a conjunction of simple and negative existential v-literals. Because of these restrictions, one can translate the effect interaction formula into a number of patterns that the tool can check for reachability. The general translation is omitted from this paper, but the following is an example of the translation, with some intermediate steps left out.

Consider the following part of an effect interaction formula:

$$\nexists v^+ : v^+ = \{p_1\} \triangleleft v \wedge v^+ = v \triangleright \{p_2\}.$$

That is, e_1 corresponds to the negative existential literal $\nexists n : v(p_1, n)$ and e_2 to $\nexists n : v(n, p_2)$. According to the general translation, this formula is found equivalent to:

$(1)\ \exists\, id' : v(p_1, id') \wedge p_2 \neq id' \vee$
$(2)\ \exists\, id : v(id, p_2) \wedge p_1 \neq id.$

This formula can finally be split into six patterns:

$$
\begin{aligned}
&\exists\ \text{different } p_1, p_2, id' : conditions(p_1, p_2) \wedge v(p_1, id')\ \text{ from } (1),\\
&\exists\ \text{different } p_1, p_2, id : conditions(p_1, p_2) \wedge v(id, p_2)\ \text{ from } (2),\\
&\exists\ \text{different } p_1, p_2 : \quad\ conditions(p_1, p_2) \wedge v(p_1, p_1)\ \text{ from } (1), id' = p_1,\\
&\exists\ \text{different } p_1, p_2 : \quad\ conditions(p_1, p_2) \wedge v(p_2, p_2)\ \text{ from } (2), id = p_2,\\
&\exists\ \text{different } p_2, id' : \quad\ conditions(p_2, p_2) \wedge v(p_2, id')\ \text{ from } (1), p_1 = p_2,\\
&\exists\ \text{different } p_1, id : \quad\ conditions(p_1, p_1) \wedge v(id, p_1)\ \text{ from } (2), p_2 = p_1.
\end{aligned}
$$

where $conditions$ is the conjunction of $allowed_1$, c_1 and c_2.

3.4 Interaction with Frame Formulas

Interaction between a frame formula (of the base specification) and a reactive rule (of a new feature) occurs if the new feature modifies the bookkeeping of the base specification. That is, for some variable v we have a frame formula

$$Changed(v) \Rightarrow \exists\, event_1 \vee \ldots \vee \exists\, event_n$$

and a reactive rule

$$
\begin{aligned}
newevent(p) &\Rightarrow allowed(x, p)\\
&\wedge\ condition(x, p) \Rightarrow v^+ = expr(v, p).
\end{aligned}
$$

In this case, there is indeed interaction if a state x is reachable in which $newevent(p)$ is allowed for such a p that the assignment to v in the above reactive rule is executed and changes v, formally

$$\exists\, p : allowed(x, p) \wedge condition(x, p) \wedge v \neq expr(v, p).$$

Using the tool in this case amounts to dead code detection: if such a state is not reachable, then the assignment $v^+ = expr(v, p)$ in the new feature is dead code. If 'no dead code' is included in our assumption that specifications are 'well-formed', then this form of interaction can be detected by a syntactic check. We briefly return to this form of interaction in Section 6.

4   Case Studies

In order to illustrate the different kinds of interaction defined in Section 3.1, two features on top of POTS are defined.

4.1   Allowedness interaction

This is the simplest kind of interaction. As an example, we take the Three-Party service (using the definition in [ITU92]), which gives an extra interpretation to the *Dialling* event of POTS.

Three-Party service   The Three-Party service enables a user who is connected to a second one to put the connection on hold; this event is named *Transfer*. Then he can dial a third party. If the third party answers, the first user can switch back and forth between the two, this is a *Switch* event. He can also make it a three-way call (all three users being connected) by another event (*3 Way*). Finally, in all these stages, there are several ways to disconnect the call (completely or in part).

We observed that the distinction between *Transfer* and *Switch* events is often not made in other papers. However, in our framework it is natural to make the distinction, as the two events differ both in allowedness and effect. The intention is to only specify *abstract* events in our framework. We assume that there is a kind of parser that receives signals from outside (e.g., *flashhook*), and delivers abstract events. Which event it delivers depends on the state of the system. We do not aim at specifying the concrete signals, and therefore we do not specify the parser in full. The allowedness conditions of abstract events can be seen as a partial specification of the parser: they specify when it is allowed to deliver the event. When two abstract events are 'bound' to the same signal, an interaction occurs if both events can be allowed at the same time. This interaction is not visible in our framework, because it occurs 'inside' the parser. We regard it as a form of resource interaction (viewing the signal as the resource) that is not analyzed in this paper. However, our framework does show that for example *Transfer* and *Switch* have mutually exclusive allowedness conditions, and can therefore be bound to the same concrete signal without risking this kind of interaction.

To give the example of allowedness interaction, we only give the relevant part of the service: the *Transfer* and *Dialling* events. *Transfer*(id) is allowed if id has no other subscriber on hold, and is not involved in a three-party call. Its conditional effects do not change the variables from POTS (logically, this follows from the frame formulas for POTS), but only put the call on hold.

$$Transfer(id) \Rightarrow$$
$$\exists\, id' : (connected(id, id') \lor connected(id', id)) \land$$
$$\nexists id'' : callhold(id, id'')$$
$$\land\ connected(id, id') \stackrel{id'}{\Rightarrow} callhold^+ = callhold \cup \{\langle id, id'\rangle\}$$
$$\land\ connected(id', id) \stackrel{id'}{\Rightarrow} callhold^+ = callhold \cup \{\langle id, id'\rangle\}$$

In the Three-Party service, dialling is more general than in POTS, because of the possibility to dial even if one is already connected. The allowedness condition for the *Dialling* event in the Three-Party service is given by the following state formula.

$$CanDialThree(id) \triangleq \left(\begin{array}{l} active(id) \\ \land\ \nexists id' : trying(id, id') \\ \land\ \neg busy(id) \\ \land\ \neg 3party(id) \\ \land\ \exists\, id' : (callhold(id, id') \lor callhold(id', id)) \end{array} \right)$$

Apart from a different allowedness condition, the reactive rule for *Dialling* in the Three-Party service is the same as the *Dialling* rule in POTS, except that it also updates the state variable *3party*.

To prove that the above is an example of allowedness interaction we show that it is possible to reach a state such that $CanDial(id)$ and $CanDialThree(id)$ are not equivalent, i.e.,

$$\exists\, id : \neg\, (CanDial(id) \Leftrightarrow CanDialThree(id))$$

holds. This is easily confirmed by the tool. There are essentially two ways to resolve the interaction. One is to replace the allowedness condition for *Dialling* in POTS by the disjunction of the two allowedness conditions $CanDial(id)$ and $CanDialThree(id)$. Another is to make *Transfer* (and *Switch*) remove the connection that goes on hold from *connected*, but that introduces an interaction between *Transfer* and the frame formulas of POTS.

4.2 Effect interaction

This is the second kind of interaction described in Section 3.1. In order to give an example, a second service is introduced.

Automatic CallBack   The Automatic CallBack (ACB) service description we use comes from [CGN+93]. The aim is that, if a connection cannot be established because the called number is busy, establishing the connection is automatically attempted again as soon as the called number is no longer busy. More precisely, we have the following scenario in mind.

1. *id* calls *id'*, but *id'* *IsBusy*, and therefore *id* becomes *busy*.

2. *id* requests the feature.

3. When eventually neither *id* nor *id'* *IsBusy*, the system tries to establish a call between *id* and *id'* by first ringing *id*.

4. When (and if) *id* answers (by picking up the receiver), he will find himself in the same state as if he had just dialled *id'*.

To illustrate effect interaction, we consider the case where *id* has requested the ACB service and his phone is ringing when *id'* is no longer busy. For this case, the specification of the ACB service uses the binary state variable *dialback* to record that *id*'s phone is ringing, and *id* has a CallBack request to *id'*. ACB also needs a binary state variable *busy2*. The case where *id* calls *id'* but *id'* is busy is recorded as $busy2(id, id')$. The rule in the ACB service that corresponds to *id* picking up his receiver is:

$$
\begin{aligned}
Offhook(id) \Rightarrow \\
\neg\ active(id) \\
\wedge \qquad\qquad\qquad\qquad\qquad active^+ &= active \cup \{id\} \\
\wedge\ dialback(id, id') \qquad\quad \xRightarrow{id'}\ dialback^+ &= dialback \setminus \{\langle id, id' \rangle\} \\
\wedge\ dialback(id, id') \wedge IsBusy(id') \xRightarrow{id'} trying^+ &= trying \wedge \\
busy^+ &= busy \cup \{id\} \wedge \\
busy2^+ &= busy \cup \{\langle id, id' \rangle\} \wedge \\
connected^+ &= connected \\
\wedge\ dialback(id, id') \wedge \neg IsBusy(id') \xRightarrow{id'} trying^+ &= trying \cup \{\langle id, id' \rangle\} \wedge \\
busy^+ &= busy \wedge \\
busy2^+ &= busy2 \wedge \\
connected^+ &= connected
\end{aligned}
$$

After the effects of this rule, the resulting state is the same as if id had just dialled id'. This gives an effect interaction: these new effects of *Offhook* are incompatible with both other effects in POTS, namely answering a call (which changes *connected*), and becoming active (which changes neither *trying* nor *busy*). Since there are two alternative effects in the ACB *OffHook* rule there are a total of four interaction formulas to check for reachability. For the case of an answering effect in POTS and the second effect in ACB (which changes *busy* and *busy2*) the interaction formula to check is:

$$
\begin{array}{ll}
& \exists\, id, id', id'' : \\
\text{allowed} & \neg\ active(id) \ \wedge \\
\text{condition 1} & trying(id', id) \ \wedge \\
\text{condition 2} & dialback(id, id'') \ \wedge\ \neg\ IsBusy(id'') \ \wedge \\
& \nexists connected^+, trying^+ : \\
\text{effect 1} & connected^+ = connected \cup \{\langle id', id\rangle\} \ \wedge \\
& trying^+ = trying \setminus \{\langle id', id\rangle\} \ \wedge \\
\text{effect 2} & connected^+ = connected \ \wedge \\
& trying^+ = trying \cup \{\langle id, id'\rangle\}
\end{array}
$$

The other cases are left to the reader. These formulas can be rewritten into patterns that can be checked with the tool.

We can also see that the new rule for *Offhook* introduces not only an effect interaction, but also an interaction with the frame formulas of POTS, in particular the one that states that *Offhook* can not change *busy*.

How can these interactions be resolved? The effect of *Offhook* in ACB must be given precedence over the *Offhook* transition in POTS that applies if $\nexists id' : trying(id', id)$. The other case, interaction between an ordinary answer of a call, and answering a *dialback*, can be resolved by adding a *dialback* disjunct to *IsBusy*. This solution is not so surprising, as *dialback* resembles *trying*. We leave the details to the reader, as interaction resolution is not the topic of this paper.

The reachability analysis tool can also be used outside interaction detection. The two state variables *busy* and *busy2* should clearly be related by the invariant $busy(id) \Leftrightarrow \exists\, id' : busy2(id, id')$ for all id. To check that this is an invariant of the system, we can check the reachability of the formula $\neg\ (busy(id) \Leftrightarrow \exists\, id' : busy2(id, id'))$. Thus we can verify *invariants* of a specification. In some cases, yet not in this example, absence of feature interaction will depend on such an invariant. Observe that the user is not required to provide the invariant in such a case: the tool can generate and verify it automatically.

5   Comparison with other frameworks

Our style of specification is inspired by the version of temporal logic presented in Lamport's TLA (Temporal Logic of Actions) [Lam91]. Intuitively, TLA is based on the specification of individual operations or actions by a relation in first-order logic that describes the state change induced by this operation, like in Z. A framework in Object-Z is outlined in [But93] and does indeed compare with our style. In addition, Object-Z has some constructs that support of building larger modular specifications, but lacks operators for describing liveness properties which temporal logic provides. To add such constructs to our formalism is a topic for future research.

While our style of specification is inspired by TLA, our suggested style for specifying telephone services is slightly more general than TLA. Similarly, in [GK94] a more general modal logic is used. However, while our linear-time temporal logic is a well known temporal logic they use a customised instance of modal logic. Compared to the earlier paper on temporal logic specifications [BJK94], we have used a more uniform framework. Apart from the advance of writing more structured service specifications it has also made it possible to make a classification of interactions.

When comparing to other approaches with a similar aim, we would like to point out the following qualities.

Properties   In several frameworks such as [CP94, RT95, EU-94] it is suggested that a specification of a service consists of a wish-list of properties wanted to be fulfilled. Assume properties P_1 for the service S_1 the correctness of S_1 can be expressed by $S_1 \models P_1$. Given these definitions a feature interaction between S_1 and S_2 occurs if $S_1 \models P_1$ and $S_2 \models P_2$, but $S_1 \oplus S_2 \not\models P_1 \wedge P_2$. That is, if each feature satisfies its properties when operating in isolation, then the failure of the composition to operate properly indicates an interaction. The exact interpretation of the composition operator $\oplus$ depends on the specification language. In our framework, the $\oplus$ is a simple conjunction, except when we need to adapt frame formulas.

We do not deal with properties as above in this paper, as our specifications are more transition-oriented. However, we can formulate such properties of the specification in temporal logic. One advantage of our approach is that we can stay in one formalism. Similarly to interaction detection as described in Section 3, a tool can be used to check for the reachability of states that do not satisfy the desired properties, with the restrictions outlined in Section 3.2.

Infinite Domains   In several other approaches, e.g., [LL94, FL94, BA94], only finite domains are considered. However, it would not be natural to limit the number of potential subscribers to some fixed finite number. A major problem when dealing with large domains is the *state space explosion* problem, which is most prominent in systems which consist of many parallel processes. Our approach, based on reasoning about patterns, is completely symbolic, and thus does not suffer from this problem.

6   Conclusions and Future Work

6.1   Handling Interaction

We have characterised feature interaction in purely logical terms, without any reference to operational semantics. Then we showed how to exploit this characterisation to detect interaction, using an existing tool. We did not address the question how to handle interaction, once detected. Deciding which of the interacting services should be altered (perhaps both) is not our concern. Our concern is how to specify the alterations in such a way that the whole specification remains within our framework: only then can our method be reapplied when further services are added.

Sometimes, an interaction can be foreseen without computation. In particular, if a new event in a feature triggers a change of an existing variable, then a frame formula is violated. If such a change is really intended (this must be verified, automatic interaction removal is not a good idea), then we must not take the conjunction of frame formulas for this variable, but their disjunction, which can be written as one frame formula.

Another situation is that two features 'share' an event. Either one feature allows the event and gives the effects of it, whereas the other disallows it, or both features allow the event, but give it different effects. In the first case one can, after verifying that the event can be allowed, remove the stronger allowedness condition. In the second case, it is likely that the event in the added feature has a precondition, the negation of which should be added to the preconditions of the base system. More systematically, one can extend the precondition of the transition that 'loses' by the negation of the preconditions of the conflicting ones that 'win'. It is also possible that neither one should win. Then the preconditions of both are extended and a conflict handler is added as a small extra 'service'. These thoughts are made more precise in [BJK94].

6.2   Refined Views of a System

Our view of a system, as depicted in Figure 1, is rather naive. It suffices for specifying requirements on the user level, but not at the network level: all users are part of a global system and the network between them is abstracted away completely! The required functionality of the network from the user's point of view could be specified in a layout depicted in Figure 2. There is still no visible network, but the users are separated: an event can only modify the status of one user; in order to modify the status of another user, a follow-up event must be generated.

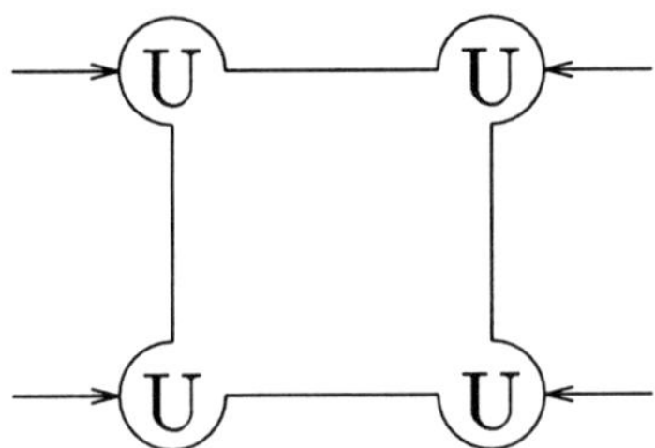

Figure 1: Our view of a system.

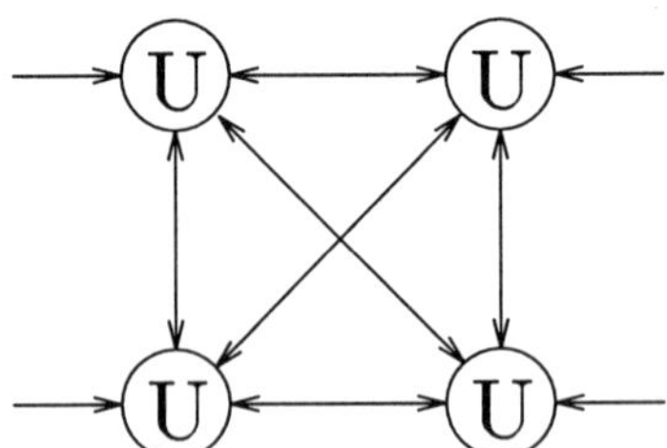

Figure 2: The user's view.

We are currently experimenting with this view of the system. For the moment, we assume (admittedly unrealistically) that follow-up events take no time, or so little time that the system stabilises before the next incoming event. This view turns out to simplify the task of writing the specification considerably! Instead of giving all effects of an event, in all cases, at once, we can separate the concerns by user.

Consider for example *Dialling*(id, id') in POTS. The event occurs at id, a case distinction is made at id', and finally the effect occurs at id again (and, in fact, also at id', namely the ringing). In the distributed view, *Dialling*(id, id') generates a follow-up event from id to id'. At id' the case distinction is made, and an appropriate folluw-up event is sent back to id, where it determines the next state. This separation of concerns is most useful when adding features: if a new feature affects only one step in the chain of follow-up events, we are not forced to rewrite everything. For example, AR affects only the case distinction.

Logically, we can view the follow-up events as natural abbreviations. (The abbreviation *IsBusy* is a rudimentary and unsystematic form of this.) By unfolding the abbreviations, we can arrive at the specifications given in this paper. Without this approach, there is little hope that the specification style of this paper scales up to realistic systems. With it, it probably does.

Dropping the assumption that follow-up events take no time requires a significant extension of our current framework. In particular, it raises the question how to deal with simultaneous events, and the meaning of 'next'. This will be a topic for further research. At first, we can retain our specifications (the ones with follow-up events), and change the underlying semantics. Then we will probably find that the specifications do not result in the intended behaviour of the features any more, and analyze the necessary changes.

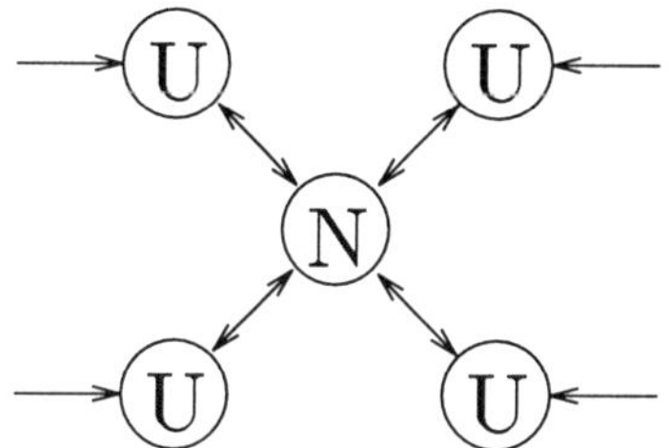

Figure 3: The global network view.

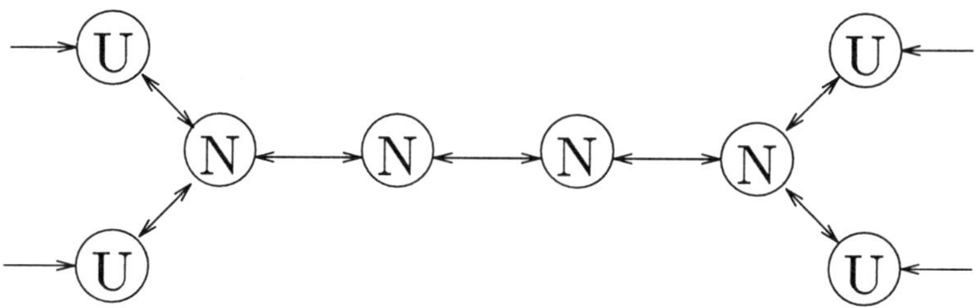

Figure 4: The distributed network view.

Returning to the different views of the system, it might turn out that the requirements for a network (in particular from a network point of view) are even easier to specify in the layout depicted in Figure 3; the network is at least present in this layout! The advantage of this layout is that we also have one 'special' entity which can maintain global state information. The functionality of the users around it might be increased gradually from simply passing through information to handling features autonomously. Whatever intermediate view turns out to be convenient, finally we must come to the level of detail depicted in Figure 4: a distributed network with users attached to various nodes.

Acknowledgement. We thank Prof. Bengt Jonsson for introducing us to this problem area and for his continuous support.

References

[AJ93]   Parosh Aziz Abdulla and Bengt Jonsson. Verifying programs with unreliable channels. In *Proc. 8th IEEE Int. Symp. on Logic in Computer Science*, 1993. Accepted for Publication in Information and Computation.

[BA94]   K.H Braithwaite and J.M Atlee. Towards automated detection of feature interactions. In L.G. Bouma and H. Velthuijsen, editors, *Feature Interactions in Telecommuniactions Systems*, pages 36–59. IOS Press, 1994.

[BJK94]   J. Blom, B. Jonsson, and L. Kempe. Using temporal logic for modular specification of telephone services. In *Feature Interactions in Telecommunications Systems*, Amsterdam, Holland, May 1994.

[But93]   M.J. Butler. Feature interaction analysis using Z. Available from ftp://ftp.abo.fi/pub/cs/papers/michael/fiz.ps.Z, 1993.

[CGN+93]   E.J. Cameron, N. Griffeth, Y.-J. Linand M.E. Nilson, W.K. Schnure, and H. Velthuijsen. A feature-interaction benchmark for IN and beyond. *IEEE Communications Magazine*, 31(3):64–69, March 1993.

[CP94]   P. Combes and S. Pickin. Formalisation of a user view of network and services for feature interaction detection. In L.G. Bouma and H. Velthuijsen, editors, *Feature Interactions in Telecommuniactions Systems*, pages 120–. IOS Press, 1994.

[EU-94]   EURESCOM Project EU-P103. Framework for service description with supporting architecture. Final report, EURESCOM, Dec. 1994.

[FL94]   M. Faci and L. Logrippo. Specifying features and analysing their interactions in a LOTOS environment. In L.G. Bouma and H. Velthuijsen, editors, *Feature Interactions in Telecommuniactions Systems*, pages 136–151. IOS Press, 1994.

[GK94]   A. Gammelgaard and J.E. Kristensen. Interaction detection, a logical approach. In L.G. Bouma, editor, *Feature Interactions in Telecommunications Systems*, pages 178–196. IOS Press, 1994.

[ITO92]   Y. Inoue, K. Takami, and T. Ohta. Method for supporting detection and elimination of feature interaction in a telecommunication system. In *Proc. International Workshop on Feature Interactions in Telecommunications Software Systems*, pages 61–81, St. Petersburg, Florida, Dec. 1992.

[ITU92]   ITU-T. Stage 2 description for multiparty supplementary services - three-party service. Recommendation Q.84, ITU, Jun. 1992.

[JK95]   Bengt Jonsson and Lars Kempe. Verifying safety properties of a class of infinite-state distributed algorithms. In *Proc. 7th Int. Conf. on Computer Aided Verification*, volume 939 of *Lecture Notes in Computer Science*, pages 42–53. Springer Verlag, 1995.

[Lam91]   L. Lamport. The temporal logic of actions. Technical report, DEC/SRC, 1991.

[LL94]   F. J. Lin and Y-J. Lin. A building block approach to detecting and resolving feature interactions. In L.G. Bouma and H. Velthuijsen, editors, *Feature Interactions in Telecommuniactions Systems*, pages 86–119. IOS Press, 1994.

[MP92]   Z. Manna and A. Pnueli. *The Temporal Logic of Reactive and Concurrent Systems*. Springer Verlag, 1992.

[OH94]   T. Ohta and Y. Harada. Classification, detection and resolution of service interactions in telecommunication services. In L.G. Bouma and H. Velthuijsen, editors, *Feature Interactions in Telecommuniactions Systems*, pages 60–72. IOS Press, 1994.

[RT95]   Tele Danmark Research and Telia. A formal approach to the detection of service interference by testing. Draft, NAUTE, Jan. 1995.

[Spi89]   J.M. Spivey. *The Z Notation*. Prentice-Hall, 1989.

Feature Interactions in Telecommunications Systems III
K.E. Cheng and T. Ohta (Eds.)
IOS Press, 1995

A Formal Framework for Feature Interaction with Emphasis on Testing*

Jens Chr. Godskesen
Tele Danmark Research
Lyngsø Allé 2, DK-2970 Hørsholm, Denmark

Abstract

In this paper we present a formal framework for feature interaction. The framework consists of three levels: *the requirement level, the specification level, and the implementation level*. At the requirement level we define interaction as inconsistency between feature requirements. At the specification and implementation level we define interaction as a *relative notion*. Interaction at the specification level is defined relative to the requirements the feature specifications are expected to satisfy. Interaction at the implementation level is defined relative to both the requirements and the specifications the implementations of the features are supposed to realize. We identify an interaction as originating from one (and only one) of the three levels. Also, we show that interactions may be inherited from the requirement level to the specification level and from the specification level to the implementation level. The final part of the paper contains an outline of how absence of feature interaction can be tested for at the implementation level.

1 Introduction

During the past few years a lot of research has been spent in the area of feature interaction. We mention the initial work reported in [3] and the work in [5] and [7, 1]. Although much effort has been spent on this problem there is still much debate as to how to deal with interaction and even on how interaction should be understood. As a contribution to this debate we propose in this paper a formal framework for *feature interaction* with the emphasis on *testing*.

Part of the research carried out contains various classifications and definitions of feature interaction. For instance, in [3] feature interaction is defined by

> "... a behaviour through which one feature inhibits or subverts the intended execution of another feature or another instance of itself, or creates joint execution dilemmas that are not resolved within the individual feature instances."

---

*Part of this work has been carried out in the Naute project, a common project between Telia Research and Tele Danmark Research. The view presented in the paper is not necessarily that of Naute.

Like in the definition above we have limited our scope of interest to that of *undesirable feature interaction.*

However, as it appears the definition above may be considered to be too inaccurate because it leaves much to be interpreted by the reader. Also, if the advantage of (semi) automatic tool support is wanted for the detection of feature interaction a more precise definition is required. In the literature one may therefore find definitions of feature interaction where formal languages has been applied. As an example feature interaction is defined as follows in [6]

> Let $F_1, \ldots, F_n$ be features. Then there is an interaction between $F_1, \ldots, F_n$ if

$$N \oplus F_1 \models P_1, \ldots, N \oplus F_n \models P_n \tag{1}$$
$$\text{but } N \oplus F_1 \oplus \ldots \oplus F_n \not\models P_1 \wedge \ldots \wedge P_n$$

where N is the specification of a network, $F_1, \ldots, F_n$ are specifications of features, and $P_1, \ldots, P_n$ are formulae expressing feature requirements.[1] $\oplus$ denotes the composition of networks and features and $\models$ denotes a satisfaction relation. Intuitively, the definition means that there is an interaction between the features $F_1, \ldots, F_n$ if each feature satisfies its requirement when composed with N, but the composition of all the features with the network N does not satisfy the conjunction of all the requirements.

In our view, but this is not stated in [6], the definition expressed by (1) makes feature interaction a *relative notion*. Not only is feature interaction defined relative to a language for expressing requirements and a language for specifying features, but interaction between the features $F_1, \ldots, F_n$ is also defined as a relative notion depending on the requirements $P_1, \ldots, P_n$. The notion is relative because it may be that $F_1, \ldots, F_n$ interact with respect to some requirements $P_1, \ldots, P_n$, whereas they do not interact with respect to other requirements $P_1', \ldots, P_n'$. In that respect the notion of interaction heavily relies on the requirements taken into account.

Further, one may argue that in (1) it is not clear whether the interaction is due to the composition of features with the network or whether the interaction is due to the conjunction of requirements. That is, it could be that the composition $N \oplus F_1 \oplus \ldots \oplus F_n$ does not satisfy the formula $P_1 \wedge \ldots \wedge P_n$ whereas $P_1 \wedge \ldots \wedge P_n$ may be satisfied by $N \oplus F_1' \oplus \ldots \oplus F_n'$ for some other features $F_1', \ldots, F_n'$. In that case it would be reasonable to argue that the interaction is caused by composition of features. However, it could also be that $P_1 \wedge \ldots \wedge P_n$ is inconsistent in the sense that it is not satisfied by any specification and in that case it would be reasonable to argue that the interaction is caused by composition of requirements and *not* by the composition of features.

This discussion suggests that one should consider interactions at the level where they rightly belong. For instance, as illustrated above the interaction may belong at the level where requirements to features are expressed or at the level where the features are specified. It would be natural to expect also that new interactions, which do not occur neither at the requirement level nor at the specification level, do occur when

---

[1] In [6] features are specified in SDL [10] and requirements are expressed using a temporal logic. For instance, a requirement (for Call Waiting) stating that a subscriber should be notified of an incoming call when already engaged can be expressed as a temporal logic formula. Intuitively one may say that a requirement defines *what* a feature must be capable of whereas a specification describes *how* the feature should behave in order to satisfy its requirements.

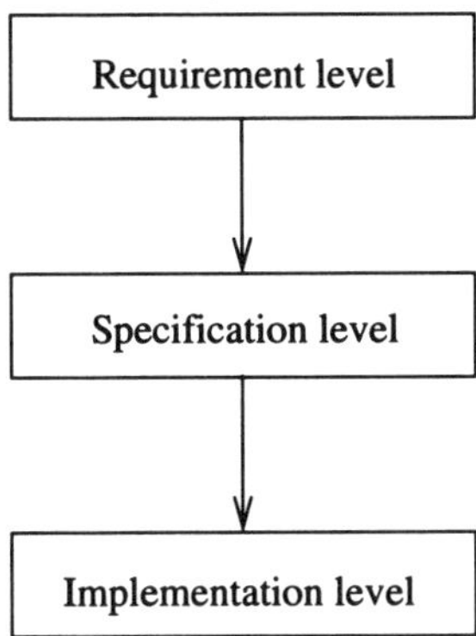

Figure 1: The three levels of interaction.

features are implemented. One reason being that a lot more details have to be taken care of.

In the proposed framework we distinguish between interaction at three distinct levels: *the requirement level, the specification level*, and *the implementation level*.[2] The three levels are shown in figure 1. The arrows in the figure illustrate that interactions may be inherited from preceding to subsequent levels.

Interactions can be detected at each of the three levels. For instance, if requirements are expressed using a logic the existence of interaction can be established by proving unsatisfiability (inconsistency). This approach has been applied in the tool Delphi [14]. At the specification level the existence of interaction can be detected by using e.g. model checking techniques. This is demonstrated in [6] with help of the tool GEODE. Also techniques known from process algebra have been used to detect feature interaction at the specification level [12]. At the implementation level it seems reasonable to apply testing techniques for detection of feature interaction. The last contribution of this paper therefore is a view on how implementations of features can be tested for absence of interaction. It follows from the inheritance of interactions that an interaction belonging to one level may be detected at subsequent levels and that an interaction can not be detected at a level preceding the one to which it belongs.

2　Feature interaction

In this section we define the framework for feature interaction consisting of the three levels of requirement, specification, and implementation. For sake of simplicity and clarity we only consider interactions between pairs of requirements, specifications, and implementations, and for the very same reason we do not consider the network component as in (1). The framework however is easily extendable to include a network component and to consider interaction between an arbitrary number of requirements,

---

[2]In [2] one may find a similar view of having levels of feature interaction. However, the notion of feature interaction is not formalized and in particular the language dependent relativity of feature interaction between levels is not addressed.

specifications, and implementations. Also, for sake of simplicity and clarity we define interaction at a rather abstract level presupposing formal languages that have to be instantiated by concrete choices.

2.1 Interaction at the requirement level

First we presuppose a *requirement language*, $\mathcal{L}_{req}$. Propositional, predicate, or temporal logics are examples of requirement languages. Also, MSC [11] may be considered a requirement language. One may view a requirement, $r \in \mathcal{L}_{req}$, as denoting all implementations satisfying r. We therefore assume an *implementation language*, $\mathcal{L}_{imp}$, and we assume an *interpretation* of requirements

$$[\ ] : \mathcal{L}_{req} \rightarrow 2^{\mathcal{L}_{imp}}$$

That is, $[r]$ denotes the set of implementations satisfying r. We straightforwardly extend the interpretation to sets of requirements, $R \subseteq \mathcal{L}_{req}$, by

$$[R] = \bigcap_{r \in R} [r]$$

Often we shall refer to R as a requirement. Intuitively R denotes the conjunction of all $r \in R$. We assume that $[\mathcal{L}_{req}] = \emptyset$, that is no implementation satisfies all requirements in $\mathcal{L}_{req}$. In the sequel we shall occasionally write $t\!t$ (true) for the empty requirement $\emptyset$, and we shall write $f\!f$ (false) for $\mathcal{L}_{imp}$.

We say that R is *inconsistent* if $[R] = \emptyset$, and we say that R is *consistent* if R is not inconsistent. It is immediate that $t\!t$ is consistent and that $f\!f$ is inconsistent.

Intuitively, we want interaction between two requirements R_1 and R_2 to mean that whenever R_1 and R_2 are viewed separately then one may find implementations satisfying R_1 and one may find implementations satisfying R_2, but when the union of all the requirements in R_1 and R_2 is considered no implementation satisfies the new requirement $R_1 \cup R_2$.

Definition 1 R_1 *interacts with* R_2 *if and only if* R_1 *and* R_2 *are consistent and* $R_1 \cup R_2$ *is inconsistent.*

We let $\mathcal{R} \subseteq 2^{\mathcal{L}_{req}} \times 2^{\mathcal{L}_{req}}$ be the set of all pairs of interacting requirements, that is

$$\mathcal{R} = \{(R_1, R_2) \mid R_1 \text{ interacts with } R_2\}$$

If $(R_1, R_2) \in \mathcal{R}$ we say that the interaction *belongs at the requirement level*.

As an example, it is immediate that the requirements $t\!t$ and R do not interact whenever R is consistent, that is $(t\!t, R) \notin \mathcal{R}$, because any implementation satisfies $t\!t$. It is also immediate that $(f\!f, R) \notin \mathcal{R}$ for any R because $f\!f$ is inconsistent.

In the project Naute[3] we have considered requirements for the features Call Waiting and Call Forward on Busy. For instance, we may let R_{CW} be a requirement to Call Waiting that intuitively denotes

---

[3]Naute is a common project between Telia Research and Tele Danmark Research. Its purpose is to adopt the ideas from conformance testing of protocols [9] to that of testing features and testing for absence of feature interaction.

> if subscriber B is already engaged and receives a new incoming call there
> should be a Call Waiting tone at B ...

and we may let R_{CFB} be a requirement to Call Forward on Busy intuitively expressing
that

> if subscriber B is already engaged and receives a new in coming call then
> the new call should be forwarded to C without acknowledged to B ...

It is quite obvious that R_{CW} and R_{CFB} are in conflict because in contrast to R_{CFB}
it follows from R_{CW} that B should be acknowledged about the new call. As a consequence it would be natural to expect that there is no implementation satisfying
$R_{CW} \cup R_{CFB}$. Hence, assuming that R_{CW} and R_{CFB} are consistent, it may be concluded
that $(R_{CW}, R_{CFB}) \in \mathcal{R}$ and that we have identified an interaction at the requirement
level.

2.2 Interaction at the specification level

We presuppose a *specification language*, $\mathcal{L}_{spec}$. As examples of $\mathcal{L}_{spec}$ we may take
languages based on state machines, like SDL [10], or languages based on process algebra,
like LOTOS [8]. We let the composition of two specifications $S_1, S_2 \in \mathcal{L}_{spec}$ be denoted
by $S_1 + S_2$.[4]

Often a specification is thought of as denoting a set of requirements. Consequently,
we assume a *satisfaction relation*

$$\mathbf{sat} \subseteq \mathcal{L}_{spec} \times \mathcal{L}_{req}$$

such that S **sat** r whenever S satisfies r. Further, we let $\mathcal{L}_{req}(S) = \{r \mid S \text{ sat } r\}$ be
the set of all requirements satisfied by S. If S satisfies all requirements in some R,
that is $R \subseteq \mathcal{L}_{req}(S)$, we write S **sat** R. Otherwise we write S **sãt** R. We assume that
S **sat** ff for no S, so no specification satisfies all requirements in $\mathcal{L}_{req}$. Also we assume
that S **sat** tt for all S.

We define interaction between two specifications S_1 and S_2 relative to two requirements R_1 and R_2. Intuitively we want that two specifications S_1 and S_2 interact with
respect to the requirements R_1 and R_2 whenever S_1 satisfies R_1, S_2 satisfies R_2, but
the composition of S_1 and S_2 does not satisfy both R_1 and R_2.

Definition 2 S_1 *interacts with* S_2 *relative to* R_1 *and* R_2 *if and only if*

$$S_1 \text{ sat } R_1, \ S_2 \text{ sat } R_2 \tag{2}$$
$$and \quad S_1 + S_2 \text{ sãt } R_1 \cup R_2$$

We let

$$\mathcal{S}_{R_1,R_2} = \{(S_1, S_2) \mid S_1 \text{ interacts with } S_2 \text{ relative to } R_1 \text{ and } R_2\}$$

If $(S_1, S_2) \in \mathcal{S}_{R_1,R_2}$, but $(R_1, R_2) \notin \mathcal{R}$, we say that the interaction *belongs at the
specification level*.

---

[4]Actually specifications may usually be composed by a whole range of operators, however for this
tentative study we let go with only one operator. In Naute features are composed using the inheritance
and specialization feasibilities of SDL.

As an example one may infer that $\mathcal{S}_{tt,tt} = \emptyset$ because tt is satisfied by any specification. Also, one may infer that $\mathcal{S}_{ff,R} = \emptyset$ for all R because S **sat** ff for no S.

It may be proved that interactions at the requirement level are inherited at the specification level in the following sense

Proposition 1 *If* $(R_1, R_2) \in \mathcal{R}$ *and* S_1 **sat** R_1, S_2 **sat** R_2 *then* $(S_1, S_2) \in \mathcal{S}_{R_1,R_2}$.

In the project Naute we have specified the features Call Waiting (S_{CW}) and Call Forward on Busy (S_{CFB}) in SDL. Assuming that S_{CW} **sat** R_{CW}, where R_{CW} is as presented above, we may detect an interaction at the specification level if we let the requirement to S_{CFB} be simply tt. The interaction occurs because it turns out that S_{CFB} neutralizes S_{CW} in $S_{CW} + S_{CFB}$ and hence the requirement R_{CW} cannot satisfied by the combined specification.

2.3 Interaction at the implementation level

Following we assume a family of *implementation relations*, $\mathbf{imp}_R \subseteq \mathcal{L}_{imp} \times \mathcal{L}_{spec}$, for any $R \subseteq \mathcal{L}_{req}$. Intuitively we want from an implementation relation that whenever I and S are related then I is a correct implementation of S and I and S satisfy the requirement R. Hence we require that

$$I \ \mathbf{imp}_R \ S \ \text{ implies } \ I \in [\![R]\!] \ \text{ and } \ S \ \mathbf{sat} \ R \tag{3}$$

If $I \ \mathbf{imp}_R \ S$ we say that I implements S relative to R. If it is not the case that $I \ \mathbf{imp}_R \ S$ we write $I \ \mathbf{i\!\!\!/mp}_R \ S$. Notice that $I \ \mathbf{imp}_{ff} \ S$ for no I and S.[5]

Letting $I_1 \| I_2$ denote the composition of implementations $I_1, I_2 \in \mathcal{L}_{imp}$ we define interaction between I_1 and I_2 relative to specifications S_1 and S_2 and relative to requirements R_1 and R_2 as follows [6]

Definition 3 I_1 *interacts with* I_2 *relative to* S_1, S_2, R_1, R_2 *if and only if*

$$I_1 \ \mathbf{imp}_{R_1} \ S_1, \ I_2 \ \mathbf{imp}_{R_2} \ S_2 \tag{4}$$
$$\text{and} \quad I_1 \| I_2 \ \mathbf{i\!\!\!/mp}_{R_1 \cup R_2} S_1 + S_2$$

That is, I_1 interacts with I_2 relative to S_1, S_2, R_1, and R_2 if I_1 implements S_1 relative to R_1, if I_2 implements S_2 relative to R_2, and if the composition of I_1 and I_2 do not implement the composition of S_1 and S_2 relative to the union of the requirements R_1 and R_2.

We let

$$\mathcal{I}_{S_1,S_2,R_1,R_2} = \{(I_1, I_2) \mid I_1 \text{ interacts with } I_2 \text{ relative to } S_1, S_2, R_1, R_2\}$$

If $(R_1, R_2) \notin \mathcal{R}$ and $(S_1, S_2) \notin \mathcal{S}_{R_1,R_2}$ but $(I_1, I_2) \in \mathcal{I}_{S_1,S_2,R_1,R_2}$ we say that the interaction *belongs at the implementation level*.

As an example, one may infer that $\mathcal{I}_{S_1,S_2,ff,R} = \emptyset$ for any S_1, S_2 and R because ff is inconsistent.

Because of (3) one may prove that interactions at the requirement level and interactions at the specification level are inherited at the implementation level in the the following sense

---

[5]A discussion of implementation relations for LOTOS can be found in [4].

[6]Like for composition of specifications, implementations may be composed in great many ways.

Proposition 2 *If $(R_1, R_2) \in \mathcal{R}$ and if $I_1 \, \mathbf{imp}_{R_1} \, S_1$, $I_2 \, \mathbf{imp}_{R_2} \, S_2$ then $(I_1, I_2) \in \mathcal{I}_{S_1, S_2, R_1, R_2}$.*

Proposition 3 *If $(S_1, S_2) \in \mathcal{S}_{R_1, R_2}$ and if $I_1 \, \mathbf{imp}_{R_1} \, S_1$, $I_2 \, \mathbf{imp}_{R_2} \, S_2$ then $(I_1, I_2) \in \mathcal{I}_{S_1, S_2, R_1, R_2}$.*

Also, one may prove that interactions can not skip a level

Proposition 4 *If $(R_1, R_2) \in \mathcal{R}$ and if $(I_1, I_2) \in \mathcal{I}_{S_1, S_2, R_1, R_2}$ then $(S_1, S_2) \in \mathcal{S}_{R_1, R_2}$.*

In Naute we expect to be able to detect interactions at the implementation level by means of testing. The ideas about testing for absence of interaction are presented in the next section.

3 Testing for absence of interaction

First a *test language*, $\mathcal{L}_{test}$, consisting of tests is presupposed. For instance $\mathcal{L}_{test}$ may be the standardized language TTCN [9]. We assume that the application of a test $t \in \mathcal{L}_{test}$ to (a running version of) an implementation I yields a *test verdict* of either **pass** or **fail**.[7] We let the application of a test to an implementation be represented by the function [8]

$$\mathbf{apply} \colon \mathcal{L}_{test} \times \mathcal{L}_{imp} \to \{\mathbf{pass}, \mathbf{fail}\}$$

and straightforwardly extend **apply** to sets of tests, $T \subseteq \mathcal{L}_{test}$, such that

$$\mathbf{apply}(T, I) = \begin{cases} \mathbf{pass}, \text{ if } \forall t \in T. \, \mathbf{apply}(t, I) = \mathbf{pass} \\ \mathbf{fail}, \text{ if } \exists t \in T. \, \mathbf{apply}(t, I) = \mathbf{fail} \end{cases}$$

We shall refer to a set of tests as a test. Notice that $\mathbf{apply}(\emptyset, I) = \mathbf{pass}$ for all I.

Naturally, we are interested only in those tests that may contribute with proper verdicts, that is, the verdict must conform with whether I actually implements S with respect to R or not. We therefore introduce the notion of a *valid test* for $I \, \mathbf{imp}_R \, S$.

Definition 4 *T is a valid test for $I \, \mathbf{imp}_R \, S$ if and only if*

$$\mathbf{apply}(T, I) = \mathbf{fail} \quad implies \quad I \, \mathbf{i\!\!\not\!mp}_R \, S$$

Hence, T is a valid test for $I \, \mathbf{imp}_R \, S$ if whenever the verdict of testing I with respect to T is **fail** then I does not implement S relative to R. If the verdict is **pass** then it may be that I implements S relative to R, at least it has not been detected by the test that the opposite is the case. Notice, that the notion of a valid test is in agreement with the statement that by testing one may only show the presence of errors, not their absence.

As an example, it is obvious that $\emptyset$ is an invalid test for $I \, \mathbf{imp}_{f\!f} \, S$ for all I and S.

In order to operate with tests that contribute with proper verdicts when testing for absence of interaction at the implementation level we would like a notion similar to validity of a test for $I \, \mathbf{imp}_R \, S$.

---

[7]Often a third test verdict, **inconclusive**, is also considered.

[8]Notice that if the implementation under test is non-deterministic then **apply** may not necessarily be a function but a relation because the application of a test may have more than one verdict.

Definition 5 *T is a valid test for $(I_1, I_2) \notin \mathcal{I}_{S_1,S_2,R_1,R_2}$ if and only if*

$$\mathbf{apply}(T, I_1 \| I_2) = \mathbf{fail} \quad implies \quad (I_1, I_2) \in \mathcal{I}_{S_1,S_2,R_1,R_2}$$

So, whenever T is a valid test for $(I_1, I_2) \notin \mathcal{I}_{S_1,S_2,R_1,R_2}$ then, if the verdict of testing $I_1 \| I_2$ with respect to T is **fail**, I_1 and I_2 do interact relative to S_1, S_2, R_1, R_2. If the verdict of the test is **pass** it may be that I_1 and I_2 interact, but the test has not been able to reveal it.

From Definition 5 it follows that $\emptyset$ is a valid test for $(I_1, I_2) \notin \mathcal{I}_{S_1,S_2,f\!f,R}$ for any I_1, I_2, S_1, S_2 and R. However, in general it may turn out to be problematic to find non trivial valid tests $T \subseteq \mathcal{L}_{test}$ for $(I_1, I_2) \notin \mathcal{I}_{S_1,S_2,R_1,R_2}$ simply because T basically must test three things, namely whether $I_1 \, \mathbf{imp}_{R_1} \, S_1$, whether $I_2 \, \mathbf{imp}_{R_2} \, S_2$, and whether $I_1 \| I_1 \, \mathbf{imp}_{R_1 \cup R_2} \, S_1 + S_1$.

In Naute we have therefore followed another and more pragmatic approach in that we consider pairs of tests, $\langle T_1, T_2 \rangle \subseteq \mathcal{L}_{test} \times \mathcal{L}_{test}$, as tests for $(I_1, I_2) \notin \mathcal{I}_{S_1,S_2,R_1,R_2}$. T_1 is assumed to be a valid test for $I_1 \, \mathbf{imp}_{R_1} \, S_1$ and T_2 is assumed to be a valid test for $I_2 \, \mathbf{imp}_{R_2} \, S_2$. The hypothesis is then that the union of the two valid tests $T_1 \cup T_2$ is a valid test for $I_1 \| I_1 \, \mathbf{imp}_{R_1 \cup R_2} \, S_1 + S_1$. If we extend the definition of **apply** by

$$\mathbf{apply}(\langle T_1, T_2 \rangle, I_1 \| I_2) = \begin{cases} \mathbf{pass}, \text{ if } \mathbf{apply}(T_1, I_1) = \mathbf{fail} \\[4pt] \mathbf{pass}, \text{ if } \mathbf{apply}(T_2, I_2) = \mathbf{fail} \\[4pt] \mathbf{pass}, \text{ if } \begin{cases} \mathbf{apply}(T_1, I_1) = \mathbf{pass} \\ \mathbf{apply}(T_2, I_2) = \mathbf{pass} \\ \mathbf{apply}(T_1 \cup T_2, I_1 \| I_2) = \mathbf{pass} \end{cases} \\[16pt] \mathbf{fail}, \text{ if } \begin{cases} \mathbf{apply}(T_1, I_1) = \mathbf{pass} \\ \mathbf{apply}(T_2, I_2) = \mathbf{pass} \\ \mathbf{apply}(T_1 \cup T_2, I_1 \| I_2) = \mathbf{fail} \end{cases} \end{cases}$$

it turns out that $\langle T_1, T_2 \rangle$ is almost a valid test for $(I_1, I_2) \notin \mathcal{I}_{S_1,S_2,R_1,R_2}$ whenever T_1 is a valid test for $I_1 \, \mathbf{imp}_{R_1} \, S_1$ and T_2 is a valid test for $I_2 \, \mathbf{imp}_{R_2} \, S_2$.

Proposition 5 *Let T_1 be a valid test for $I_1 \, \mathbf{imp}_{R_1} \, S_1$ and let T_2 be a valid test for $I_2 \, \mathbf{imp}_{R_2} \, S_2$ then, if $T_1 \cup T_2$ is a valid test for $I_1 \| I_2 \, \mathbf{imp}_{R_1 \cup R_2} \, S_1 + S_2$,*

$$\mathbf{apply}(\langle T_1, T_2 \rangle, I_1 \| I_2) = \mathbf{fail}$$
$$implies \quad (I_1, I_2) \in \mathcal{I}_{S_1,S_2,R_1,R_2}, \; I_1 \, \mathbf{i\!\!/mp}_{R_1} \, S_1, \; or \; I_2 \, \mathbf{i\!\!/mp}_{R_2} \, S_2$$

That is, if $\mathbf{apply}(\langle T_1, T_2 \rangle, I_1 \| I_2)$ gives **fail** as verdict then either there is an interaction or one of I_1 and I_2 is not a correct implementation. The reason why $\langle T_1, T_2 \rangle$ is not a proper valid test is that the verdict **pass**, when applying the tests T_1 and T_2, does not imply that the implementations I_1 and I_2 are correct, it only indicates that an error has not been found. However, since $\mathbf{apply}(\langle T_1, T_2 \rangle, I_1 \| I_2) = \mathbf{fail}$ it must be that I_1 has passed the test T_1 and that I_2 has passed the test T_2, so as far as T_1 and T_2 are concerned it is reasonable to believe that $I_1 \, \mathbf{imp}_{R_1} \, S_1$ and that $I_2 \, \mathbf{imp}_{R_2} \, S_2$ and hence that $\langle T_1, T_2 \rangle$ is a valid test for $(I_1, I_2) \notin \mathcal{I}_{S_1,S_2,R_1,R_2}$.

In conclusion, we propose to test for absence of feature interaction by first testing the individual features by valid tests T_1 and T_2. If at least one of the verdicts by the outcome of the tests T_1 and T_2 are **fail** there is no interaction. Otherwise, if both verdicts are **pass** the combination of the features should be tested with $T_1 \cup T_2$. If the verdict of that test is also **pass** no interaction has been detected by the test, but if it is **fail** it may be that there is an interaction; at least one of the implementations I_1, I_2, and $I_1\|I_2$ are incorrect.

In Naute we have been able to generate tests T_{CW} and T_{CFB} for the SDL specifications S_{CW} and S_{CFB} respectively using the SDT tool [13]. We have detected an interaction at the specification level by executing the test $T_{CW} \cup T_{CFB}$ against $S_{CW} + S_{CFB}$. As explained in Section 2 the interaction occurs because S_{CW} is neutralized in $S_{CW} + S_{CFB}$. We expect to be able to find the same interaction at the implementation level for the combination of the implementations of the two features.

4 Conclusion

In this paper we have described a formal framework for feature interaction consisting of the three levels of requirement, specification, and implementation. Interaction at the specification level is defined relative to requirements posed on the features and interaction at the implementation level is defined relative to both these requirements and to specifications of the features. We have shown how interactions may be identified as belonging to one and only one of the three levels. Also, it has been shown that interactions may be inherited at lower levels. Finally, we have outlined an approach for testing for absence of interaction at the implementation level. The approach advocates to test the combination of the feature implementations with the test for all the implementations of the individual features.

In order to achieve a clear and concise presentation of our work in only ten pages the framework has been presented in a rather abstract fashion. It is the hope although that our ideas will be matured and used in practice. At least we believe they may provide as input to a better understanding of the substantial problem of feature interaction – an understanding which perhaps is crucial in order to obtain better practical solutions. Currently the ideas are utilized in the Naute project.

References

[1] L.G. Bouma and H. Velthuijsen, editors. *Feature Interactions in Telecommunications Systems*. IOS Press, 1994.

[2] T.F. Bowen, F.S. Dworak, C.-H. Chow, N.D. Griffeth, G.E. Herman, and Y.-J. Lin. Views on the feature interaction problem. Technical Report TM–ARH–012849, Bellcore, 1988.

[3] T.F. Bowen, F.S. Dworak, C.-H. Chow, N.D. Griffeth, G.E. Herman, and Y.-J. Lin. The feature interaction problem in telecommunication systems. In *Proceedings of the Seventh International Conference on Software Engineering for Telecommunications Swithching Systems*, Bournemouth, United Kingdom, 1989.

[4] E. Brinksma, G. Scollo, and C. Steenbergen. LOTOS specifications, their implementations and their tests. In *Proceedings of the Sixth International IFIP Symposium on Protocol Specification, Testing and Verification*. Elsevier Science Publishers B.V., 1987.

[5] E.J. Cameron, N. Griffeth, Y.-J. Lin, M.E. Nilson, and W.K. Schnure. A feature-interaction benchmark for IN and beyond. *IEEE Communications Magazine*, 31(3), march 1993.

[6] P. Combes and S. Pickin. Formalization of a user view of network and services for feature interaction detection. In L.G. Bouma and H. Velthuijsen, editors, *Feature Interactions in Telecommunications Systems*. IOS Press, 1994.

[7] *International Workshop on Feature Interactions in Telecommunications Software Systems*, St. Petersburg, Florida, USA, 1992. IEEE Communications Society.

[8] ISO. *Information Processing Systems, Open Systems Interconnection, LOTOS — A Formal Description Technique Based on the Temporal Ordering of Observational Behaviour*, 1989.

[9] ISO. *Information Technology, Open Systems Interconnection, Conformance Testing Methodology and Framwork*. ISO, 1991. International Standard IS–9646.

[10] ITU. *Z.100: Specification and Description Language (SDL)*, 1994.

[11] ITU–TS. *ITU Recommendation Z.120: Message Sequence Chart (MSC)*. ITU, ITU General Secretariat - Sales Section, Place des Nations, CH-1211 Geneva 20, 1992.

[12] B. Stepien and L. Logrippo. Feature interaction detection using backwards reasoning with LOTOS. In *Proceedings of the Fourtienth International IFIP Symposium on Protocol Specification, Testing and Verification*, June 1994.

[13] Telelogic AB. *SDT 3.0 – Getting Started*. Telelogic AB, Box 4128, S-203 12 Malmö, Sweden, 1994.

[14] F. Widebäck and G. Stålmark. Definition av Delphi (in swedish). Technical Report NP/FW/001, Logikkonsult NP AB.

Issues of Non-Monotonicity in Feature-Interaction Detection

Hugo VELTHUIJSEN
KPN Research, St. Paulusstraat 4, 2264 XZ, Leidschendam, The Netherlands
h.velthuijsen@research.ptt.nl

Abstract. Techniques for verifying that a system satisfies certain properties have achieved wide support as promising tools for detecting interactions between features in a telecommunications system. However, until now these techniques have not been able to detect on a large scale interactions that were not already known to exist. Surely, some interactions have been found, but any careful analysis of feature descriptions such as necessary for the formulation of feature specifications and properties that can be fed into a verification tool would most likely lead to the detection of some interactions. Part of the reason why detection mechanisms based on verification techniques have not been able to go beyond mere 'proof-of-concept' examples seems to lie with the fact that extension of a basic telecommunications system with new features is essentially a non-monotonic operation: the telecommunications system itself is modified to accommodate the new features. The purpose of this paper is to identify where non-monotonicity presents obstacles for the successful application of verification techniques to interaction detection (three separate issues are described). Identification of these issues and obstacles will hopefully lead to a re-focusing of research efforts in this area. Some pointers to formal techniques that deal with non-monotonicity are included.

1. Introduction

Probably the most active area related to the problem of feature interactions is the use of Formal Description Techniques (FDTs) for the detection of interactions. Most of the contributions in this area use formal program verification techniques for checking that a program or system satisfies certain properties. In mathematical notation:

$$S \models \phi,$$

where S is a mathematical representation of the behavior of the system and ϕ is a property, i.e., a proposition about the behavior of the system. The latter description is sometimes termed a behavioral description, while the first is sometimes termed procedural description.

Based on such a mechanism, the premise is that we can detect certain feature interactions in the following way. Given descriptions of the network N and two features S_1

reference	system language	property language
[2]	temporal logic	temporal logic
[3, 4]	LOTOS	LOTOS
[5, 12]	LOTOS	LOTOS
[6]	FSM[1]	state-data constraints
[7]	transition systems	state-transition constraints
[8]	LOTOS	LOTOS
[9, 10]	SDL	state-data constraints
[13]	transition systems	state-transition constraints
[15]	Object-Z	Object-Z
[17]	Promela (FSM)	temporal logic
[19]	FSM	state-transition constraints

Table 1: Some detection approaches based on formal methods.

and S_2, and properties ϕ_1 (for S_1) and ϕ_2 (for S_2), and assuming $N \oplus S_1 \models \phi_1$ and $N \oplus S_2 \models \phi_2$, there exists an interaction when $(N \oplus S_1) \oplus S_2 \not\models \phi_1 \wedge \phi_2$. We use here the operator $\oplus$ to signify combination of system descriptions. The actual semantics of this operator is not clearly defined and will be the topic of further discussion later in this paper.

Within the general concepts of the basic approach to interaction detection as described here, there are still several choices to be made, such as which formalism to use for describing the system, which formalism to use for describing properties, which mechanism to use for verification, and how or at what particular level of detail the system description should be modeled.

Several of these choices have been or are being tried in various research activities. Table 1 illustrates some of the alternatives found in the literature. Still others are proposed regularly. In a way, it is understandable that such a plethora of alternative, competing approaches within the same general framework are proposed, because there is no common understanding to what particular choices are best for detecting certain or all feature interactions.

However, in my mind diversification in this vein is currently besides the point. The techniques used so far have failed to detect on a large scale interactions that were not known to exist beforehand. None of the approaches have progressed far beyond a mere proof-of-concept result. The approaches show that with the right descriptions and techniques it is indeed possible to detect certain interactions. But the trick is to come up with the right descriptions: the right network and service specifications, and above all, the right properties. This appears very hard to do well and presents for the time being a more urgent challenge than the application of yet another specification formalism to the same general approach.

In this paper, I point out several obstacles that were encountered while taking the same general FDT-based approach in the RACE project SCORE [20]. As I will show in this paper, part of the reason that these obstacles exist lies with the fact that adding new features to a telecommunications service leads in many cases to non-monotonic extensions. In brief, adding a new feature can often not be done without modifying the

---

[1]FSMs (finite-state machines) are used regularly, with certain slight differences in their semantics, which we do not discuss here.

original service. This non-monotonicity turns up as an obstacle in at least three different situations when applying FDTs to interaction detection. These three situations may not be causally unrelated, but become an issue at separate stages of the FDT-based approaches. The three issues are discussed in turn in Sections 2, 3 and 4.

Classical logics (Predicate Logic, First Order Logic) and temporal logics do not address non-monotonicity. Yet, these logics are at the heart of the specification languages and verification tools developed in the area of FDTs and used for current efforts to detect feature interactions. This suggests that these languages and tools are currently inadequate to deal effectively with the feature-interaction problem.

Non-monotonicity has been the focus of extensive study in the area of Artificial Intelligence (e.g., [14]). The purpose of this paper is to identify the challenges that are posed to detection approaches due to the non-monotonicity of adding features and to provide some hints to existing results that may serve to extend current FDTs to make them more suited to dealing with the feature-interaction problem.

2. Non-monotonic extensions

Services and features are extensions to a telecommunications system. In the basic interaction-detection approach, a procedural description (i.e., specification or design) of the telecommunications network is assumed. This network description can be extended by descriptions of the features. Furthermore, the approach requires descriptions of features in connection with supporting network functionality to act as criteria of good behavior for the verification process. The procedural and behavioral descriptions form together a formal model of the network and features.

There is also another model at play here that is usually not made explicit. Feature developers do not have a complete description of the network in mind when designing a new feature but rather a mental model of the network and features. This mental model contains not only factual information about the network and features, but also statements *about* the network and features that the developer knows to be true. These statements can be seen as (informal) properties of the network and the features. My conjecture is that this mental model is used to a large extent by feature developers as a basis for designing and implementing new features. The different models are shown in Figure 1.

Extension of a network specification is often denoted with the infamous O-plus operator ($\oplus$). The semantics of this operator is always left unspecified, and for good reason: the addition of a feature to a telecommunications system, and consequently the modeling thereof, is not straightforward. One of the reasons for this is that such extensions in the metal and formal models are usually non-monotonic extensions: the extension necessitates modifications of the original network and feature models used by feature developers and in interaction analysis. Most features change some of what was previously held to be true in the mental model of the original telecommunications system. This can be illustrated by any number of examples:

- The **Freephone** service makes it possible that the callee is charged for a call instead of the caller.

- The **Call-Waiting** feature introduces different versions of the busy state: the original one where a called party is busy and the caller hears a busy tone; and a new one where the called party may be busy, but the caller hears a ringing tone.

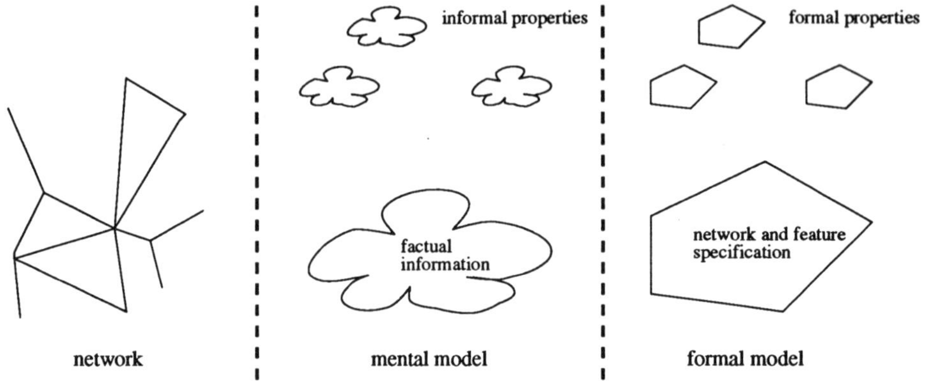

Figure 1: Different models of network and features.

- The **Call-Forwarding** feature makes that the dialed number and the number of the line to which the caller is connected are no longer synonymous.

- **Call Screening** features make that dialing a number will no longer automatically lead to a call set-up attempt.

- and so on...

Market forces necessitate the ability to create new services independently from each other. The number of new services as well as the number of different organizations involved in creating and deploying new services makes it infeasible that one service developer knows about all the different ways in which other services and features modify the telecommunications system and the mental and formal models of the system the developer uses. Thus, a service developer will assume the network model as he perceives it and base a new service on this model. However, other services that are created independently may very well violate one or more of the assumptions made in the mental network model. This presents an important source of interactions.

Thus, certain interactions may be introduced due to the non-monotonicity of mental models when adding new features. The same non-monotonicity occurs in the models that are used in interaction-detection analysis and hinder adequate detection. This non-monotonicity is one of the reasons that the O-plus operator is rather elusive. Let us examine why the O-plus operator has so far escaped formal treatment in publications about detection approaches. Part of this elusiveness lies with the fact that even if certain service functionality presents a perfectly monotonic extension to the network model, the semantics of this extension may vary. For example, in some cases an extension may be modeled adequately by the addition of new states to an otherwise constant finite state machine; in other cases an extension is more suitably modeled by the addition of a concurrent process. Another problem with the O-plus operator is that it does not seem to be commutative.

Non-monotonicity further complicates the semantics of the O-plus operator in the following way. Let N be a model of a telecommunications network and S_1 and S_2 models of two services. Extension of the network with the new service S_1 is denoted as

$N \oplus S_1$, which could be re-written as $N' + S_1$, where $+$ is some 'monotonic' extension operator, and N' is a modified network model. Modification of N has become necessary to accommodate the new service S_1. Similarly, $N \oplus S_2$ becomes $N'' + S_2$, where N'' is another modified network model. But how should we rewrite the formulae $(N' + S_1) \oplus S_2$ and $(N'' + S_2) \oplus S_1$? Knowing this involves knowing how the network model modifications necessitated by the service S_2 affect the model of S_1 and vice versa, which means knowing interactions between the two services even before we are in a position to use our detection mechanism!

This observation suggests that non-monotonicity needs to be incorporated in some way in the specifications of the telecommunications network. However, the formalisms that are used in current FDTs are based on classical logics such as Predicate Logic (PL) and First Order Logic (FOL) and these formalisms are strictly monotonic. This means that if we know $P_1 \wedge \ldots \wedge P_n \to Q$, then it must also be true that $P \wedge P_1 \wedge \ldots \wedge P_n \to Q$ (the addition of the new fact P does not negate the previously valid implication).

One may wonder if anything can be done to formalize non-monotonicity in order to create FDTs that are better equipped to express this issue and incorporate it into the models and the analysis techniques used for interaction detection. The following provides some pointers.

In the area of Artificial Intelligence (AI) it has been noted that monotonicity does not hold for some forms of (often quite useful) human reasoning. Examples of this are:

hypothetical reasoning: a mechanism to explore consequences of different hypotheses about the world (what-if scenarios); and

default reasoning: the adoption of certain propositions which may be falsified later when new information becomes available, but are necessary not to become completely paralyzed.

In these situations, it may very well occur that $P \wedge P_1 \wedge \ldots \wedge P_n \not\to Q$, even though $P_1 \wedge \ldots \wedge P_n \to Q$. Such forms of reasoning have led within AI to the development of formalisms that deal with non-monotonicity, such as Reiter's Default Logic (a logic that provides a formal basis for the inference of retractable information, [21]), and de Kleer's Assumption-based Truth Maintenance System (a system that keeps track of inferences and the assumptions they are based on in order to be able to retract them when the assumptions are violated, [11]). Further study is necessary to understand if and how results from non-monotonic reasoning in AI can be used to augment efficiently and effectively current FDTs. Non-monotonic reasoning techniques are notoriously inefficient, but quite a bit of progress has been made in this area over the last few years, especially for restricted applications.

3. Formulating properties

In formal interaction-detection approaches, properties are used to describe the required behavior of a service. A service specification or implementation is checked against these properties to see if there is an interaction. Although this should be evident, I want to stress that an interaction is found *only* if the service behavior violates its required properties. Thus, the properties are the only criteria to determine with this approach whether an interaction exists. Consequently, this places considerable importance on the formulation of adequate properties. But it has proven to be extremely difficult

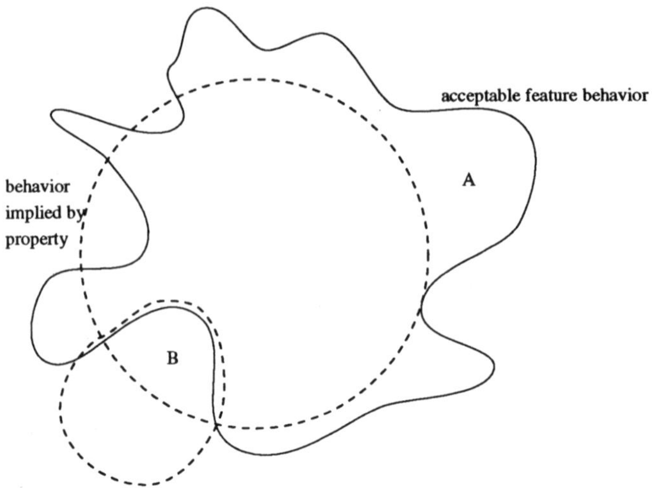

Figure 2: Covering acceptable behavior of a service by formulating properties.

to formulate properties that help to detect interactions that were not known to exist beforehand.

A property[2] should not be less strict than the actually acceptable behavior of a service. Because in that case combined services may exhibit unacceptable behavior that is not detectable by verifying this property. Alternatively, a property should not be overly restrictive, because then interactions may be detected that are actually no interactions at all. In practice, such a situation would still be useful, since manual verification might very well be capable of separating the 'real' interaction cases from the rest. Then, the automated detection method serves as a method for cutting down the number of cases (feature pairs) that need to be checked manually. However, if too many 'false' interactions are detected by an automatic approach, the value of the automatic approach deteriorates quickly. Thus, the success of an automatic detection mechanisms depends on good coverage of acceptable behavior by the properties. But how do we achieve good coverage?

The idea of coverage is illustrated in Figure 2. Experience with formulating properties tells us that it is not so hard to come up with a first approximation (the circle in Figure 2). Further thinking about a service's behavior helps us to identify where the initial approximation was too restrictive or not restrictive enough (areas A and B, respectively) and provide pointers for improvement. But in what directions do we have to think to come up with improvements? Because the properties must help to decide if there is an interaction between two features, it follows that good directions for finding improvements are provided by how the feature should behave in the presence of other features. Regrettably, such an exercise seems to involve doing a manual interaction analysis as a prerequisite to do automatic interaction analysis. This rather defeats the purpose of doing the automatic analysis in the first place!

Furthermore, a perfect coverage does not seem to be feasible. Unless we can find a

---

[2] A set of properties that are intuitively seen as covering different characteristics of a service can in logical terms be seen as a single property by the use of conjunction.

way to conclusively describe a service's behavior through the use of some (declarative) property language, the service description in terms of properties will always be an abstraction of a service design or implementation and hence not cover all of its required behavior.

Consequently, if properties cannot cover all of the required behavior of a service, we have to choose which aspects we will describe and which aspects we won't. However, what are good criteria for choosing? Again, good criteria are typically those that tell us how a service should behave in combination with other services. And again we are required to do a manual interaction analysis, just to be able to do it automatically. I illustrate this with properties described in [18].

Consider the feature Call Forwarding Unconditional (CFU). This feature can be described in a first approximation by the following:[3]

$$AG(A \neq B \wedge A \neq C \wedge CFU(A, B)) \wedge ready(C) \Rightarrow [dial(C, A)]calling(C, B))$$

This formula states that if A has used CFU to forward calls to B, then dialing A's number by C results in C calling B. However, when we consider this property in relationship with the Basic Call Service, we find that this property does not fully cover the required behavior. For instance, it does not deal adequately with the fact that B may be either idle, or busy. Thus, we need to modify the property in the following way:

$$AG(A \neq B \wedge A \neq C \wedge CFU(A, B)) \wedge ready(C) \wedge idle(B) \Rightarrow [dial(C, A)]calling(C, B)) \bigwedge$$

$$AG(A \neq B \wedge A \neq C \wedge CFU(A, B)) \wedge ready(C) \wedge busy(B) \Rightarrow [dial(C, A)]rejecting(C, B))$$

Of course, this is still not quite right if we take into account that user B may subscribe to Call Waiting (CW), which means that the predicate $busy(B)$ needs to be further elaborated. Note that CFU and CW do not necessarily interact, so even though we have to precizise the property due to the existence of another feature, we have not detected an interaction as a result of that. But we do detect a potential interaction when we consider how the property for CFU should be modified to represent correct behavior if we allow for user B to be subscribed Terminating Call Screening (TCS). Because then we have to determine if it is okay that C can be calling B, even though C might be on B's screening list, and that A can forward calls to B even though A might be on B's screening list.

Partly, the same non-monotonicity of the previous section raises its ugly head here again. Many aspects are not relevant and need not be mentioned in the property of a feature, since these aspects are not covered by the network model at the time that the feature is developed, but are introduced only later when new features are introduced that make these aspects an issue.

This last observation suggests two related partial solutions. These solutions would require a thorough investigation of all services and features that can currently be envisaged (the hundreds of feature descriptions contained in [1] might be a good starting point) to make a temporarily complete list of relevant aspects.[4] This list could then be used to create a network model that covers each of the aspects in the list. The two

---

[3]The notation and semantics are from the branching time temporal logic ACTL*. Details are given in [18]. Here it suffices to know that the modality AG denotes 'in all states along all time paths' and the operator $[a]$ denotes 'inevitably after a'.

[4]Such a list would most likely be temporary. Whenever such a list is created, there is nothing to stop someone to say: "Wouldn't it be neat if we had a service that could do such and such?", and by doing so introduce a new concept and new aspects to be considered.

solutions differ from the fact whether the new model would actually be used to develop a new service platform, or just as a mental model to form a checklist for writing properties that cover 'all' relevant aspects. The latter approach can be applied for existing service platforms, such as IN. The former requires the development of a new service platform and could be included in current developments such as TINA.

4. The qualification problem

The field of AI has studied extensively theories for representing and reasoning about action. Especially the domain of planning has driven a substantial research effort in this area. Planning is the task of constructing a partially ordered set of primitive operators that will transform a given begin state into a required end state. Evidently, this is an important capability for systems exhibiting any kind of 'intelligent', purposeful behavior, such as robots. For reasoning purposes, primitive operators are described in terms of preconditions on the state in which an operator can be applied and a formulation of how the resulting state differs from the begin state. This description abstracts from the actual implementation of the actions performed by an operator. Thus, the relationship between the description of an operator used for planning and the implementation of the operator itself is similar to the relationship between a property of a feature and the design or implementation of the feature.

In fact, the kind of issues that are represented in the descriptions of planning operators are similar to the issues that need to be expressed by formal properties of features. In both cases, one needs to express what will be the effect of certain activities. In the case of planning these effects are the resulting world state after applying an operator; in the case of feature properties, these effects are the telecommunications system state that occurs as a result of certain user actions and the system behavior as it should behave with the addition of a certain feature. As an illustration of such a property, remember the property for Call Forwarding Unconditional (CFU) in the previous section:

$$AG(A \neq B \wedge A \neq C \wedge CFU(A, B)) \wedge ready(C) \Rightarrow [dial(C, A)]calling(C, B))$$

This property provides a description what the effect $(calling(C, B))$ should be of an action $(dial(C, A))$ on future system states, given a begin state (A has activated CFU to B; C is in a 'ready' state).

Formalization of the principles underlying planning, which is necessary for automation of the process, has shown that classical logics such as Predicate Logic or First-Order Logic (FOL) are inadequate to deal with two problems, termed the *qualification problem* and the *frame problem*, respectively.[5]

The qualification problem stems from the fact that the successful execution of an operator may depend on a very large number of preconditions (termed qualifications), too many to list explicitly and many more than are usually even considered to be relevant.

For instance, consider the feature Automatic CallBack (ACB; invoking the ACB feature should lead to a call attempt to the line from which a call was received last). Successful application of ACB requires that certain conditions are fulfilled. Rather obvious may be that the user invoking ACB is subscribed to the feature and that someone has previously called the user. Already more farfetched conditions are that

---

[5]Some publications do not distinguish between the qualification problem and the frame problem and denote both with the single term 'frame problem'.

the ACB user is not on the call screening list of the person who called previously and that that person has not forwarded calls to his line to another line.[6] Even more farfetched are conditions such as that the ACB user's phone should be jacked in and that his local switch is not down.

The qualification problem introduces non-monotonicity to formulation of properties. Imagine that we have a situation where the effect (E) of a feature is thought to depend on a certain precondition (P): $P \rightarrow E$. Then, for instance because a new feature is introduced, we find out that the effect also depends on another precondition (P_n) and we find that in fact $\neg P_n \wedge P \nrightarrow E$, which contradicts monotonicity.

An example of such a situation in writing properties for features is the following. We consider the feature Calling Number Display (CND) and its relation to the displaying of private, unlisted numbers. An intuitive formulation of the fact that a private number is not displayed by CND is this:

$$private(A) \rightarrow call(A, B) \wedge \neg display(A)$$

However, when we take into account an emergency service such as the North American 911, this condition is overruled, and we have:

$$private(A) \wedge (B = 911) \nrightarrow call(A, B) \wedge \neg display(A)$$

Apart from the number of preconditions that may be too overwhelming for enumeration, the number of aspects that are *not* affected by the execution of an operator may become very large; also too large for explicit enumeration. Yet, the description of what stays the same during the execution of an action is necessary in planning in order to infer reliably that certain facts are still true in a new state.

It is not clear to me if this form of the frame problem actually presents a problem to the detection of feature interactions. Detection mechanisms are primarily used to find out if a resulting system state conforms to an explicitly stated requirement. Explicit enumeration of certain aspects in the resulting system state that are *not* affected by a certain feature does not seem to be very useful, since these aspect would almost by definition not be the topic of any interactions.

The frame problem might be an issue, however, if we try to formulate concisely how a feature modifies an existing service description and try to use this formulation to generate automatically a new and complete service description. Enumeration of all those aspects of the original service description that are not modified by the addition of a new feature is not feasible, since we do not know beforehand all these aspects as they may be added only later by the introduction of new features. Consequently, we would not be able to derive a complete new description if we do not address this version of the frame problem in some way.

The field of AI has produced a number of techniques to deal with the qualification and frame problems, most notably circumscription, which is a non-monotonic extension to FOL. Circumscription formalizes the notion of the Closed World Assumption (CWA) which is used regularly in classical planning systems to deal with the frame problem and states that any negated predicate is true unless the unnegated form of the predicate is explicitly given. See for a more extensive discussion of these issues, e.g., [16].

---

[6]Of course, these two conditions are suggested by interactions that exist between ACB on the one hand, and Terminating Call Screening (TCS) or Call Forwarding Unconditional (CFU) on the other hand.

5. Conclusions

In this paper, I argued that detecting feature interactions using Formal Description Techniques (FDTs) has to deal with non-monotonicity. I showed three different situations where non-monotonicity presents obstacles to the successful application of many currently applied FDTs:

Non-monotonic extensions: The addition of new features to an existing telecommunications service requires in many cases modifications of the original service (both of the real service and of the models used for detection of interactions).

Property (re)formulation: The formulation of properties in such a way that they cover adequately allowed system behavior requires more or less manual interaction analysis and often needs to be re-done when the addition of new features introduces new aspects that need to be covered by the properties.

The qualification problem: The formulation of properties has to deal with the qualification problem as it also occurs in AI planning systems.

This is not to say that these three different situations are in fact independent. Each might actually be causally traced back to the fact that adding features to a telecommunications system usually requires non-monotonic extensions. But this issue does turn up in different forms during the various activities that constitute a feature-interaction detection method based on FDTs and lead to circular arguments in a number of cases.

One way to deal with non-monotonicity would be to develop a service platform that supports the addition of most if not all features only through monotonic extensions. Given that the definition of such a service platform seems to be utopian for the time being, more results may be expected from improving current interaction detection approaches.

In order to be able to deal with issues of non-monotonicity and to fulfill the expectations that many researchers have for the usefulness of FDTs for detecting feature interactions, current FDTs need to be improved. The field of AI has studied non-monotonicity extensively and developed formalisms for dealing with it. An interesting topic for further study would be the investigation if and how these results from AI can be used to overcome the obstacles that non-monotonicity presents to the successful application of FDTs to feature-interaction detection.

Acknowledgements

The ideas presented in this paper have largely been fueled by the bottlenecks encountered in the RACE project SCORE [20], although the ideas themselves were developed outside this project. However, I am indebted to the KPN Research SCORE team for providing me with the material that proved instrumental in the development of the ideas presented in this paper. I further thank the SCORE team, i.e., Wiet Bouma, Willem Levelt, Alfo Melisse, and Kees Middelburg, for their discussions which helped me to clarify these ideas.

References

[1] LATA switching systems generic rquirements (LSSGR). Report TR-TSY-000064, FSD 00-00-0100, Bellcore, July 1989.

[2] J. Blom, B. Jonsson, and L. Kempe. Using temporal logic for modular specification of telephone services. In L.G. Bouma and H. Velthuijsen, editors, *Feature Interactions in Telecommunications Systems*, pages 197–216, Amsterdam, May 1994. IOS Press.

[3] L.G. Bouma and J. Zuidweg. Towards formal analysis of feature interactions (abstract). In *Proceedings International Workshop on Feature Interactions in Telecommunications Software Systems*, page 156, St. Petersburg, FL, 2-4 December 1992.

[4] L.G. Bouma and J. Zuidweg. Formal analysis of service/feature interaction using model checking. Report TI-PU-93-868, PTT Research, Leidschendam, The Netherlands, February 1993.

[5] R. Boumezbeur and L. Logrippo. Specifying telephone systems in LOTOS. *IEEE Communications Magazine*, 31(8):38–45, August 1993.

[6] K.H. Braithwaite and J.M. Atlee. Towards automated detection of feature interactions. In L.G. Bouma and H. Velthuijsen, editors, *Feature Interactions in Telecommunications Systems*, pages 36–59, Amsterdam, May 1994. IOS Press.

[7] E.J. Cameron and Y.-J. Lin. A real-time transition model for analyzing behavioral compatibility of telecommunications services. In *Proceedings ACM SIGSOFT '91*, 1991.

[8] K.E. Cheng. Towards a formal model for incremental service specification and interaction management support. In L.G. Bouma and H. Velthuijsen, editors, *Feature Interactions in Telecommunications Systems*, pages 152–166, Amsterdam, May 1994. IOS Press.

[9] P. Combes and S. Pickin. Formalisation of a user view of network and services for feature interaction detection. In L.G. Bouma and H. Velthuijsen, editors, *Feature Interactions in Telecommunications Systems*, pages 120–135, Amsterdam, May 1994. IOS Press.

[10] P. Combes, B. Renard, W. Bouma, and H. Velthuijsen. Formalisation of properties for feature interaction detection. In *Proceedings RACE IS&N Conference*, pages III/3/1–19, Paris, 23-25 November 1993.

[11] J. de Kleer. An assumption-based tms. *Artificial Intelligence*, 28(2):127–162, 1986.

[12] M. Faci and L. Logrippo. Specifying features and analysing their interactions in a lotos environment. In L.G. Bouma and H. Velthuijsen, editors, *Feature Interactions in Telecommunications Systems*, pages 136–151, Amsterdam, May 1994. IOS Press.

[13] A. Gammelgaard and J.E. Kristensen. Interaction detection, a logical approach. In L.G. Bouma and H. Velthuijsen, editors, *Feature Interactions in Telecommunications Systems*, pages 178–196, Amsterdam, May 1994. IOS Press.

[14] M.L. Ginsberg, editor. *Readings in Nonmonotonic Reasoning*. Morgan Kaufmann Publishers, Inc., Los Altos, CA, 1987.

[15] A. Lee. Formal specification — a key to service interaction analysis. In *Proceedings of the Eighth International Conference on Software Engineering for Telecommunications Software Systems (SETSS '92)*, Florence, March/April 1992.

[16] V. Lifschitz. *Formal theories of action*, pages 410–432. Morgan Kaufmann Publishers, Inc., Los Altos, CA, 1987.

[17] F.J. Lin and Y.-J. Lin. A building block approach to detecting and resolving feature interactions. In L.G. Bouma and H. Velthuijsen, editors, *Feature Interactions in Telecommunications Systems*, pages 86–119, Amsterdam, May 1994. IOS Press.

[18] C.A. Middelburg. A simple language for expressing properties of telecommunication services and features. Report PU-94-356, PTT Research, Leidschendam, The Netherlands, April 1994.

[19] T. Ohta and Y. Harada. Classification, detection and resolution of service interactions in telecommunications systems. In L.G. Bouma and H. Velthuijsen, editors, *Feature Interactions in Telecommunications Systems*, pages 60–72, Amsterdam, May 1994. IOS Press.

[20] RACE project 2017 (SCORE). Report on the methods and tools for service creation (second version). Volume 1: service interaction analysis. Technical Report SCORE Deliverable D204, 1994.

[21] R. Reiter. A logic for default reasoning. *Artificial Intelligence*, 13(1):81–132, 1980.

Runtime Resolution of Feature Interactions in Architectures with Separated Call and Feature Control [1]

Norbert Fritsche
Technische Universität München, Lehrstuhl für Kommunikationsnetze
Arcisstr. 21, D–80290 München, Germany
norbert@lkn.e–technik.tu–muenchen.de
WWW: http://www.lkn.e–technik.tu–muenchen.de/persons/nf_e.html

Abstract. The major obstacle to achieve a rapid deployment of features is the handling of the interactions between the increasing number of features. This encompasses the flexible adaptation of the feature interaction behavior corresponding to customer requirements, as well as the prevention of errors and feature execution conflicts.

The feature interactions are classified in three groups. First, functional interactions represent the specified interaction behavior between features. Second, event interactions occur because of coexisting feature instances, which are able to process the same event. The third class is made up by error interactions. These arise because of architectural and modelling constraints.

This paper presents multiple techniques to handle the different kinds of interactions. These techniques are combined to a resulting algorithm to resolve feature interactions. Features are described through multiple feature roles. These feature roles can be taken by devices during the feature life cycle, if all feature roles of one device are compatible with each other. Next, a generic feature state machine is introduced. This state machine evaluates the desired state transition of a feature before an event will be assigned to a specific feature instance. This allows to distinguish critical or uncritical state transitions corresponding to feature interactions. The next part of the combined feature interaction algorithm is to define general rules for the event distribution. To prevent error situations between interacting features, the required actions are included in the feature role mechanism when the role compatibility is checked. The resulting algorithm resolves feature interactions at runtime.

An architecture with separated call and feature control and 11 features has been implemented and tested in prototype form by using a SDL tool. The presented feature interaction algorithm, which resolves the interactions between these 11 features (> 80 specified functional interactions) has been developed and tested.

1. Introduction

The main goals of new switching architectures are, firstly, to provide new flexibility and, secondly, to enable rapid feature creation, as well as the modification of existing features. These goals will be reached by separating call control and feature control (eg. Intelligent Networks IN [1]). To attain a strict separation of the call control and feature control, the switching system is reduced to its basic capabilities and the whole feature control is concentrated into a functional unit, communicating via a protocol with the basic switch. The basic switch performs

---

1 This research is funded by Siemens AG, Private Communication Systems, Munich, Germany

the basic call processing and provides via a call model informations about the actual call processing, reports events to the feature control and provides a mechanism to influence calls and connections. The result is, consequently, a well–structured modular system architecture. The separation can be "virtual", as in an internal architecture (switching software architecture), as well as physical, as in IN.

Because the presented feature interaction examples are described by the notation of the used call model and some feature interactions are the result of modelling constraints, the used call model will be introduced shortly [2]. The basic call (2–party call) (Fig.1) consists of two devices (eg. terminal, trunk), with one connection each. The connection with its correlated connection state machine represents the relationship between device and call. The connections are linked together by the call object. These three objects, call, connection and device, are all call model objects defined within the call model. The same state machine is applied for originating and terminating side of a call. An IN–like event generation mechanism is provided for the connection. These events provide information about a change in state of a connection. The call model provides half–calls. A half–call consists of a device with one connection and a call object, without a second connection. In the following figures devices are illustrated as rectangles,

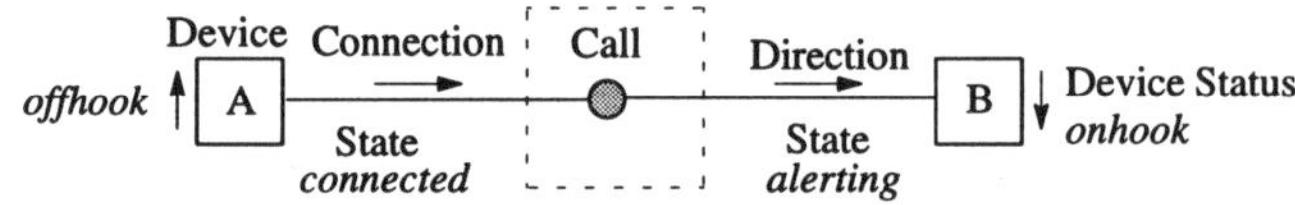

Fig.1: Call model

the relationship of a device to a call – a connection – as a line. A call, which links together two connections, is represented as a circle.

For the architectures with separated call and feature control the major unsolved problem is to handle the feature interactions. This encompasses the avoidance, detection and resolution of interactions. The terms service and feature are defined in [3], [4].

<u>Feature interactions</u> are understood to be all interactions that interfere with the desired operation of the feature and that occur between a feature and its environment including other features or other instances of the same feature. Additionally, interference of one part of a feature with another part of that feature is also considered to be a feature interaction [3]. A feature interaction can be a desired interaction, when features are designed to interact with another, as well as undesired interactions, where error situations (eg. deadlocks) occur. It is also important to ensure required non–interaction. Example: *Call Forwarding Unconditional* (CFU) – *Terminating Call Screening* (TCS) (Fig.2)

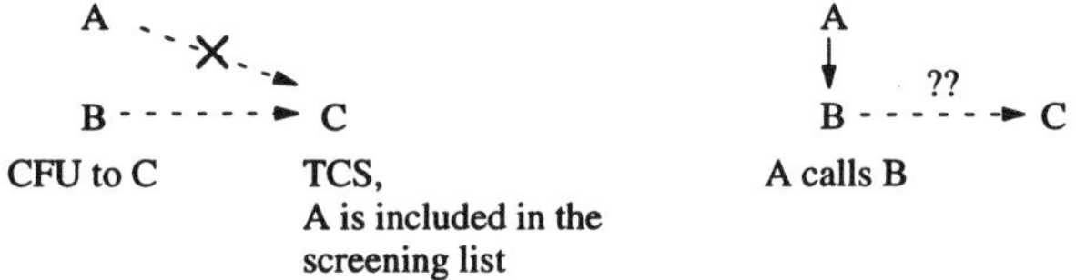

Fig.2:　Feature interaction between *Call Forwarding Unconditional* (CFU) and *Terminating Call Screening* (TCS)

B forwards its incoming calls to C, but C has activated TCS and has put A to its screening list. When A calls B, should the call be forwarded, and if so, should it be screened by C ? When is it possible to resolve these interactions? There are multiple possibilities to resolve these interactions, when activating CFU or TCS, or when CFU will be invoked, or when the forwarded call is presented to C and the TCS will be invoked.

In the context of this paper, a feature interaction (FI) algorithm is a method to systematically resolve all or most of the feature interactions.

The <u>requirements</u> of a feature interaction algorithm are as follows. Basically the feature interaction handling should be separated from the different features. It is very important, that the feature interaction algorithm is feature independent, so that new features can be added without any change to the existing logic. The interactions should be brought into the system as data, which specify the interaction behavior between the new and the existing features. This requirement is very important for private switching systems to enable an easy customization of the interaction behavior between specific features. The effort for adding features should be low. Additionally, the feature interaction concept should provide systematic support for the whole development of new features and their interactions, beginning with the specification. It should support generic interactions (eg. blocking of feature activations), as well as detailed interactions, so that it is possible to enter specific feature interaction treatment procedures (eg. a maximal number of sequenced call forwardings).

The general feature interaction problem and the corresponding responsibility of the feature interaction manager (FIM) in architectures with separated call and feature control consists of the following aspects:

<u>Correct distribution of switch events to the interested features (Fig. 3):</u>

In general, there are multiple features interested in the same event. These features can use this event in different ways:

- *Activating*:　　A feature can be activated by using the event, eg. activating *Call Forwarding*.
- *Initiating*:　　An activated feature can be invoked by using the event, eg. set up of a re–call.
- *Subsequent*:　An already invoked feature (active feature instance) can be continued by using the event, eg. the partner of an *Automatic Recall on Busy Subscriber* (ARBS) call goes offhook.

Before the FIM distributes an event to the correct feature, the FIM has to determine all features which are interested in an occurred event. It is also possible to use an event by multiple features simultaneously.

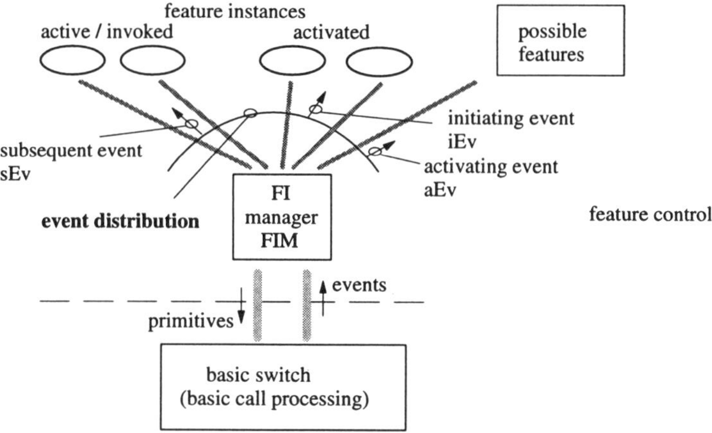

Fig. 3: Responsibilities of a Feature Interaction Manager (FIM)

<u>Compatibility</u>:

The already activated or active (invoked) features and the newly activated or invoked features have to be compatible. Before the FIM is able to perform the compatibility check, it has to determine all interacting features (see section 2).

<u>Feature specification</u>:

The FIM has to control the feature interactions in that way, that the customer specifications will be met. These specifications describe the desired behavior of specific interacting features.

<u>Coexistence of feature instances</u>:
After checking the compatibility between features, the FIM has to take some actions to enable
the simultaneous existence of features and its instances. These actions prevent error situations.

In [5] the feature interactions are classified corresponding to the phases of the feature design
in which the interaction occurs. Additionally, a feature design phase can add feature interactions, which are not visible in the former phase. In this paper three classes of feature interactions are distinguished, which are caused by the functional specification, the system architecture, the modelling constraints (eg. call model) and the feature implementation. These classes
are:

- Functional interactions
- Event interactions
- Error interactions

The <u>functional feature interactions</u> arise because of the functional behavior of the different
features. The way of interworking is specified (customer requirements) to ensure desired and
to prevent undesired interactions. These interactions are visible to the user. Problems arise because of incomplete or incorrect feature specifications. Example: *Forwarding* of an *Automatic
Recall on Busy Subscriber.*

The <u>event interactions</u> and the error interactions are caused by the switching architecture
with separated call and feature control, when mapping the functional feature control flows to
the architecture. Features consume events, which they request explicitly from the basic switch
(event interface feature control – basic switch), so that different feature instances are able to
process the same event. These interactions, which are generally not visible, when a feature is
specified functionally, are called event interactions. It is possible that two features are functionally compatible, but there can still exist event interactions between these features.

Example: The initiator A of a *Consultation Call* releases the *Consultation Call* by going
onhook (Fig.4). Terminal D has already activated an *Automatic Recall on Busy Subscriber*
(ARBS) of A. What should be done first – to offer the hold call to A to take it again or to create
the *Automatic Recall* ?

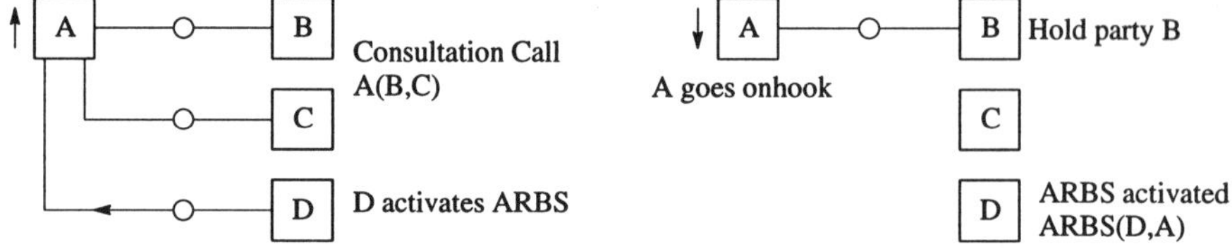

Fig.4: Event interaction *Automatic Recall* (ARBS) – *Consultation Call* (CC)

The <u>error interactions</u> occur because of the constraints of the architecture and the call model
[2]. These interactions become apparent in logical problems within the features and in deadlock situations.

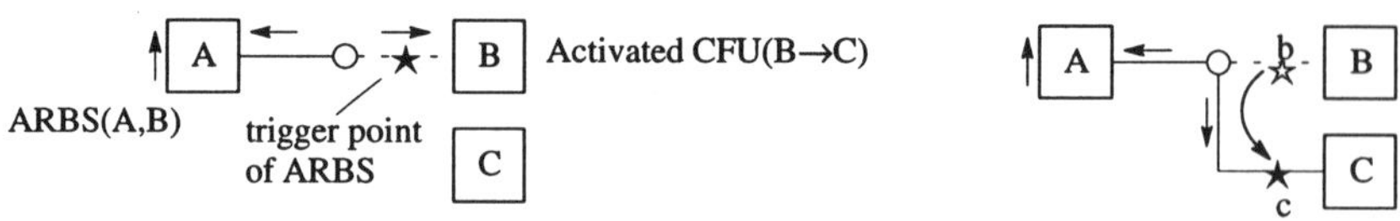

Fig.5: Error interaction *Automatic Recall on Busy Subscriber* (ARBS)
– *Call Forwarding Unconditional* (CFU)

Example: When the partner of an activated *Automatic Recall on Busy Subscriber* (ARBS)
has activated *Call Forwarding Unconditional* (CFU), the ARBS will be forwarded (if this is
the specified functional behavior) (see Fig.5). Because the call model provides only connection
events (contain no information about the complete call), the ARBS has to wait for the event:

'B accepts incoming ARBS call'. This event of connection b will not be generated, if the incoming ARBS call will be forwarded to device C. The result is a deadlock situation of the ARBS instance. To resolve this error situation, the requested event has to be generated by the new connection c.

Corresponding to the state of a feature, two different kinds of interactions occur: the <u>activation</u> and the <u>invocation</u> feature interaction (FI).

- activation FI: One feature is newly activated, others are still activated/active.

- invocation FI: One feature is newly invoked (occurrence of its initiating event), others are still activated or active.

The treatment of the interaction between two features depends on the feature states. In general, it is possible to influence both features, eg. to block an activation, to block an invocation or to deactivate an activated feature. If a feature is already invoked, the possibilities of influencing this feature instance are restricted. To avoid deadlock situations, an invoked feature always has to be terminated regularly.

There are only a few approaches to resolve feature interactions at runtime. In [1], [8] or [9] generic rules for the distribution of events to interested features are proposed. The precedence is defined by feature characteristics, eg. persistence or feature priority. Through this approach a specified interworking of features cannot be obtained. In addition, no mechanisms are provided to prevent error situations.

Schessel [7] presents a table oriented approach to resolve feature interactions. The algorithm determines all features, which have to be triggered, when an event occurs. For this, data tables are used containing all features in priority order. This approach is not sufficient for the problem of resolving interactions. It does not consider the parameters of a feature instance. In addition, it is very difficult to build a priority list of all features.

In [6] conflicts between features are resolved at runtime by an automatic negotiating system. The users have to define policies representing the constraints for feature interactions. To achieve a correct interaction, the involved user agents have to negotiate a proposal, which is acceptable for all of them.

The remainder of this paper is organized as follows. Section 2 presents a new mechanism to resolve functional feature interactions, the term *feature role* is introduced and the technique of using feature roles to handle functional feature interactions is described. The error interactions and the required actions to avoid these error situations are presented in Section 3. Section 4 explains the developed *generic feature state machine*. Event interactions and some general rules to handle these interactions are proposed in section 5. The resulting algorithm to resolve feature interactions, which is built up of the three former mechanisms, is presented in section 6. Additionally the results which have been achieved with the algorithm is described. The last section 7 explains the realization of the feature interaction algorithm, its effect on the feature design and the architecture of the developed prototype.

2. Functional Feature Interactions

2.1. Feature Roles

Some features only affect one device and its related connections, some affect devices of a single call and some can change the behaviors of multiple related calls. As a result the interactions can arise in a variety of ways. The problem of checking the compatibility between different features is very difficult. For example, if one feature is newly activated, it is necessary to determine all interfering features. In the example in figure 6 the newly activated feature is F_5

and the interacting features can be F_1, F_2, F_3, as well as F_4, when the functionality of F_5 associates in the future device D in its control flow. But how is it possible to locate all these features?

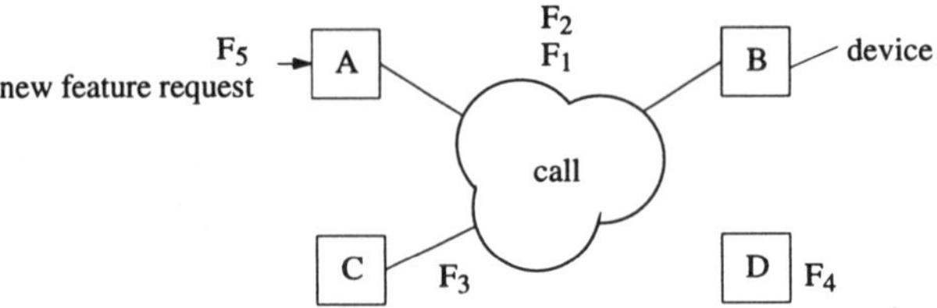

Fig. 6: Distribution of features – locating interacting features

For this purpose it is necessary to define a reference point, where the association of the device with a specific feature instance can be determined. It is proposed here to define the device as a reference point. But it is not enough to know all devices, which are affected by a specific feature control flow. For the interactions it is necessary to know how a feature acts on the calls and connections of a device. To describe the way of acting on the connections and calls of a device the term *feature role* is introduced. The feature is described through multiple feature roles, one for each distinguished relationship.

The relationship between feature, feature instance, feature roles and devices is illustrated in figure 7. A feature controls different devices (A, B, C) with its related calls and connections. These devices take feature roles during the feature control, which represent the control relationship to the feature (A–role_{i1}, B–role_{i3}, C–role_{i2}).

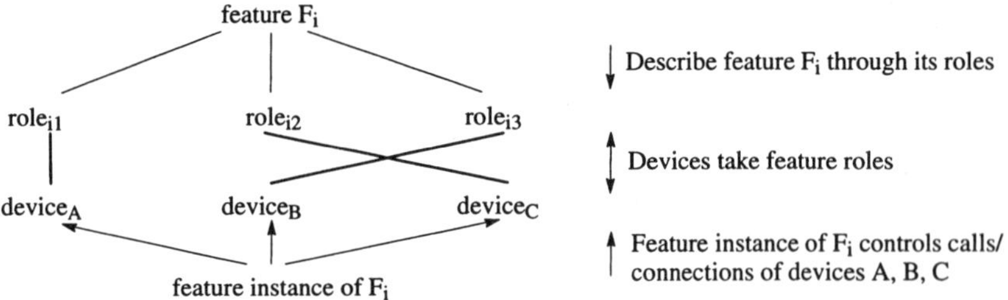

Fig.7: Relationship between feature, feature instance, feature roles and devices

Devices take feature roles during the feature life cycle. These roles are associated with the control of a specific feature instance in a feature dependent manner. The taking and the release of roles is coupled with events, which are important for the feature. These events specify the duration of life of the different feature roles. It is possible to include parameters to the feature roles, which determine the relationship to the affected call or to partner roles of the same feature.

The feature roles have to be specified for each feature, but can also be grouped for similar features (eg. all possible *Forwardings*). The parameters for all feature roles have to be determined. The differentiation between static feature roles (activation of a feature, which waits for its invoking event), dynamic feature roles (invocation of a feature) and active roles (features without an activation procedure, eg. *Consultation Call*) have to be done. To determine the life cycle of the feature roles, the events which are responsible for their taking and release have to be defined.

Figure 8 shows the roles with their actual parameters of the feature *Call Forwarding Unconditional* (CFU). The devices B and C take static roles, when CFU is activated. When the incoming call generates the initiating event for the invocation of the activated feature, the dynamic roles are taken in addition.

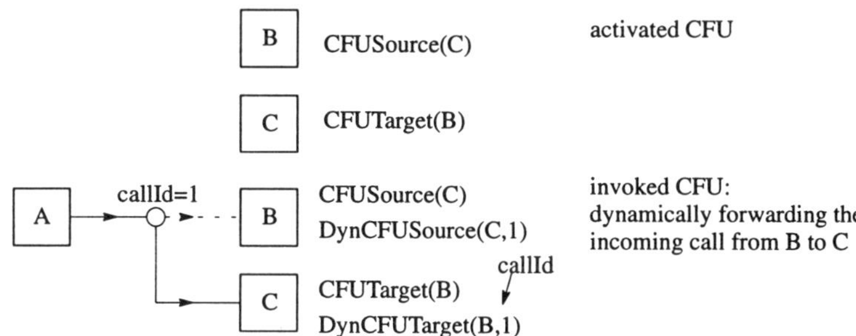

Fig. 8: Feature roles of *Call Forwarding Unconditional* (CFU)

2.2. Using Feature Roles to Resolve Functional Feature Interactions

The devices take the feature roles in the different states of the feature control. To ensure a correct functional behavior of the system (and of its interacting features) the already taken roles of one device have to be compatible to newly requested roles for this device. Hence follows, that the functional interaction between two features is mapped to a compatibility check between the roles taken and the newly requested role for each of the affected devices. In this way the complex problem of feature interactions is broken down into multiple easier problems of handling roles. Figure 9 shows the mapping of an interaction between two features F_i and F_j to the check of the two roles for device A: the earlier taken feature $role_{i3}$ and the newly requested $role_{j1}$.

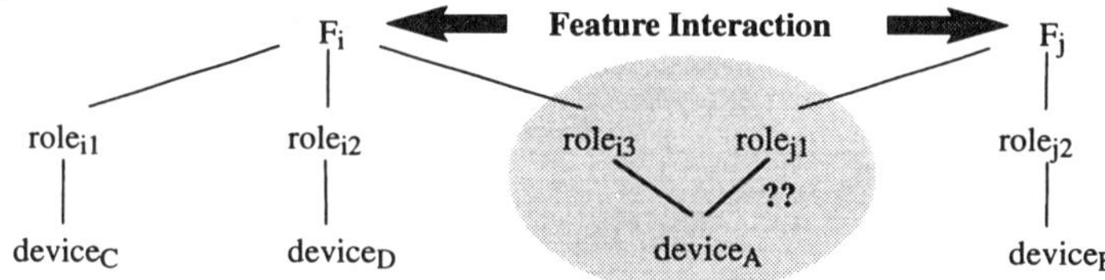

Fig.9: Functional feature interactions and the compatibility of feature roles

The user demands different feature interaction behaviors for the feature activation and the feature invocation. Additionally the FIM has different possibilities to resolve the interactions. These possible actions on the feature control flow depend on, whether a feature is invoked or whether it has to be activated or invoked. This distinction between activation and invocation will be enabled by the request of static feature roles at the activation and the request of dynamic feature roles at the invocation.

Figure 10 shows an example of the feature role mechanism. An *Automatic Recall On Busy Subscriber* (ARBS) is activated between A and B, whereby A is the initiator. B has forwarded his calls to C (CFU B→C). All corresponding static roles are already taken by the devices (Fig.10 a)). When B becomes idle the ARBS feature will be invoked. This is related with the taking of the dynamic ARBS role for A (*DynARBSInitiator*). After A accepts the incoming ARBS call (half call), the dynamic ARBS role for B (*DynARBSPartner*) is requested and the connection to B is added (Fig.10 b)). It is assumed that these two feature roles can be taken without performing any actions. When setting up the connection to B, the initiating event for the invocation of the activated CFU occurs. This is related with the request of all dynamic feature roles of CFU (*DynCFUSource, DynCFUTarget*) and the compatibility check with the existing roles for each device (Fig.10 c)). The customer has to provide information about the desired behavior of the two interacting features (Specification: should the recall be forwarded or not) to enable the checking of the two roles *DynCFUSource* and *DynARBSPartner* for the device B and to achieve this desired interaction behavior. If the ARBS should be forwarded,

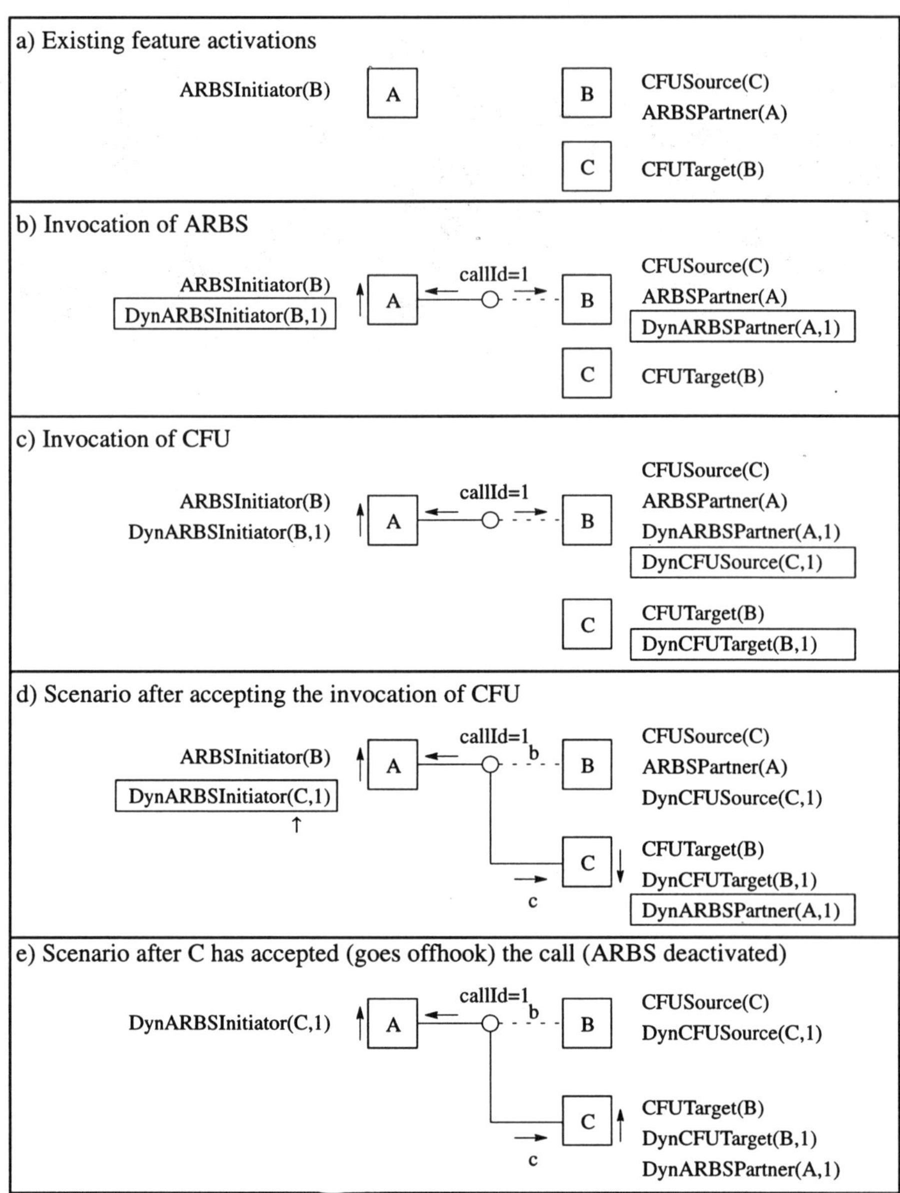

Fig.10: Example *Automatic Recall on Busy Subscriber* (ARBS) and
Call Forwarding Unconditional (CFU)

the two feature roles are compatible, but there are some actions to be performed. The dynamic roles of the forwarded call have to be taken over to C, so that the dynamic partner of the ARBS is C. Additionally the device parameter of the dynamic partner role (*DynARBSInitiator*) has to be adapted. The ARBS feature is waiting for the event, that the ARBS partner accepts the call (the corresponding detection point is set). But if the call is forwarded to another destination this detection point has to be transformed to the new destination with its connection to the call (prevent deadlock for the feature ARBS). These two actions (taking over dynamic roles and transformation of detection points) have to be performed to ensure the coexistence of the two features CFU and ARBS. Fig.10 d) shows the resulting situation. When C goes offhook and the ARBS call is successfully established, the ARBS will be deactivated and the corresponding static roles have to be cleared (Fig.10 e)). The dynamic ARBS roles and the dynamic CFU roles will be cleared when A or C releases the call.

Devices take dynamic roles when a feature is invoked (eg. incoming call for CFU source, invocation of *Consultation Call*). The related call is important for all dynamic roles, because it is required to determine if a specific call is under control of a feature (eg. incoming call is *Conference* setup) and has to be handled in a different way than an ordinary two party call. For this it is necessary to provide the call identification as a parameter of dynamic roles.

The feature role mechanism enables a flexible resolution policy. The interactions can be resolved at different points in the feature control flow of two interacting features. The resolution of the interactions between the two features CFU and ARBS can be done, when activating ARBS, when invoking ARBS or when CFU will be invoked. But the resulting interaction behavior will be quite different. In the presented example (Fig.10) the interaction is resolved when CFU is invoked for the ARBS partner device.

2.3. Matrix of Feature Roles

The principle of roles enables the check of the compatibility of two roles, one still taken and the second newly requested. If a device has already taken multiple feature roles, all taken roles will be checked in pairs with the newly requested feature role. The required information to check the roles can be stored in form of a matrix (Fig.11). When a new feature is added, the elements of the role matrix have to be determined and the feature designer has to specify the compatibility with all existing feature roles.

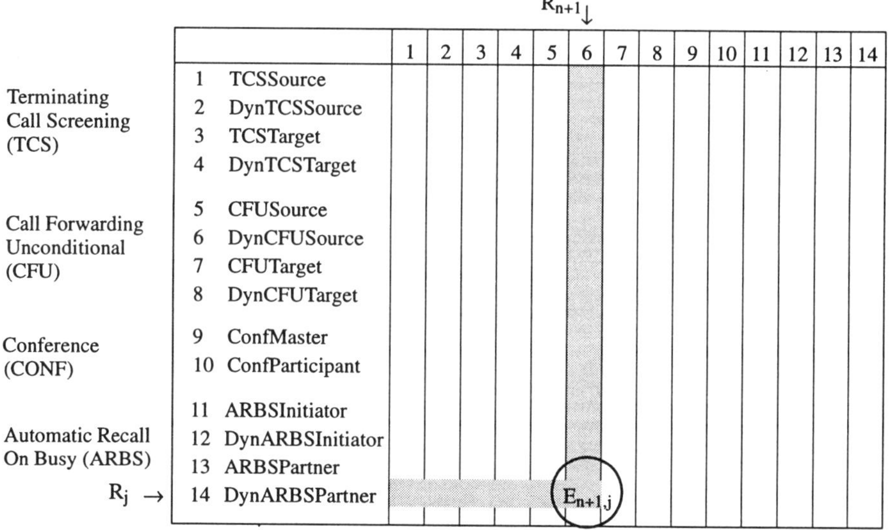

Fig.11: Part of the feature role matrix

The elements $E_{n+1,j} = (C, CC, CP, (A_1, ..., A_k))$ of the matrix for the check of the taken role R_j and the requested role R_{n+1} contain 4 different kinds of information:
- <u>Compatibility (C):</u> no
  yes   general
        parameter dependent

- <u>Algorithm to check the compatibility (CC)</u>
  In general, it is not sufficient to check the two roles without considering its parameters. This information defines the check algorithm and the parameters (CP) of the two feature roles to be checked (eg. to compare the call identifications of the two dynamic roles). In addition, the parameters of the corresponding feature instances (eg. feature state, feature priority) can be checked to determine the compatibility.

- <u>Actions to be performed to achieve the desired functional interaction behavior (A)</u>
  - Reject new role $R_{n+1} \Rightarrow$ blocking of the activation or invocation of the corresponding feature.
  - Deactivate one of the feature instances
  - Override activation parameters
  - Release of the taken role R_j
  - Transform / combine roles to a new one
  - Send display messages to devices
  - Enter a special treatment procedure

- <u>Action to ensure the consistency of roles (A)</u>
  - Take over of dynamic roles

Figure 11 illustrates the checking of the requested role R_{n+1} *DynCFUSource* against the existing role R_j *DynARBSPartner* (corresponds to Fig.10 c)). If the specified behavior of these interacting features is to block the invocation of CFU, the matrix element is defined as (a) and if the forwarding of the ARBS call is allowed, the element is defined as (b):

	(a)	(b)
Compatibility:	parameter dependent	parameter dependent
Check:	parameters: call id	parameters: call id
	==	==
Actions:	reject new role	accept new role
		take over R_j (*DynARBSPartner*) to parameter1 of R_{n+1}, the target of the CFU

This means, when the ARBS call is forwarded (call identification of the ARBS call is equal with the call identification of the requested CFU role), then the request of the feature role for the invocation of the CFU will be rejected in case (a) or accepted in case (b). In (a) the forwarding will be blocked, in (b) the call will be forwarded and the dynamic role of ARBS will be taken over to the target of the CFU.

The newly requested role is checked in pairs with all previously taken roles of the device. If the new role is compatible with all others, the actions which are specified in the matrix are performed and the new role is taken, but if it is incompatible the new role is rejected and the feature cannot be activated or invoked.

3. Error Interactions

The error interactions between compatible features arise because of the architectural and modelling constraints and especially the used call model. For the developed architecture with separated call and feature control 4 kinds of error situations are detected:

- Deadlock situations
- Logical feature problems, because of missing partner device information
- Invalid static feature activation parameters
- Not sufficient information for devices in interaction cases

In section 1 and section 2.2. the deadlock situation between *Automatic Recall on Busy Subscriber* (ARBS) and *Call Forwarding Unconditional* (CFU) has been described. This error can be prevented by transforming the corresponding triggers to the connection of the target device of the activated CFU.

Another case of error occurs between ARBS and *Terminating Call Screening* (TCS). The call model of the basic switch provides the capability to create half calls. For a half call exists no valid partner information in the basic switch. Additionally, for a conference call with multiple connections in one call, there is no valid partner information provided by the switch. Figure 12 shows the situation between *Automatic Recall on Busy Subscriber* (ARBS) and *Terminating Call Screening* (TCS) with the corresponding taken feature roles when the error situation occurs. There is an activated ARBS between device A and B, in which A is the activator of the ARBS (*ARBSInitiator*). In addition a TCS has been activated by device A and device C has been included in the screening list. When ARBS is invoked, the half call to device A is created and the initiating event for the activated **TCS feature** (incoming call for A) is generated. TCS is invoked and the feature instance attempts to **determine** the call partner. This partner information is required to analyze, whether the partner (caller) is included in the screening list. But the partner information cannot be determined **by** accessing the call model object data of the basic switch, because there is no valid partner information for the created half call. To resolve this feature execution conflict, the FIM has to **evaluate** the call partner (which will be the future call partner of the ARBS call) by the examination of the actual feature roles.

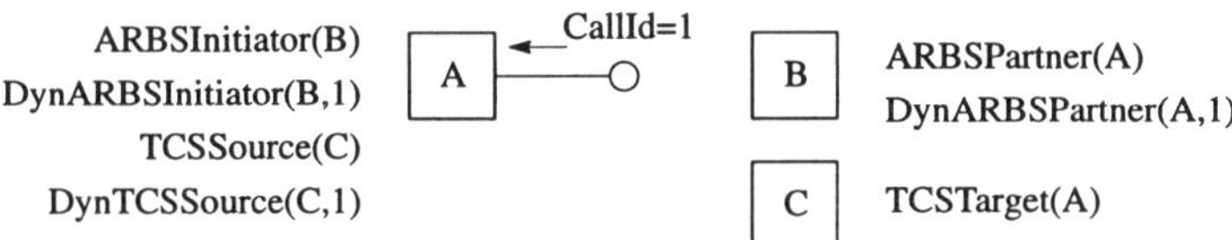

Fig.12: Missing partner information: *Automatic Recall* (ARBS) – *Terminating Call Screening* (TCS)

The feature control stores the activation parameters of the existing feature instances in its database. In the interaction example of section 2.2. the ARBS call is forwarded by the *ARBSPartner*. When the ARBS feature instance sends a display text to the actual *ARBSPartner* to inform it of the ARBS call, the feature instance cannot use these activation parameters. The actual call partner (*DynARBSPartner*) has been changed and is different to the stored information. To resolve this conflict the ARBS instance has the possibility to access the dynamic roles of the ARBS feature instance to determine the actual partner. The feature role mechanism takes care of the role consistency, so that the required information will be available, when the feature instance accesses the dynamic feature roles in contrast to the static activation parameters.

To inform the users of specific interaction situations, the delivery of display texts to the affected devices is required.

The actions to prevent error situations because of feature interactions can be included in the role matrix as actions which will be performed, when a feature role is requested and taken. Following actions have to be provided:

- Transformation of trigger (detection) points

- Determination of actual or future call partner information

- Sending display messages to devices

The actual call partner will be determined when the feature requests the corresponding feature role (if required). The call partner information will be transmitted in the acknowledgement of the feature role request to the feature instance.

4. Generic Feature State Machine (GFSM)

The generic feature state machine (GFSM) defines all points in the life cycle (or feature control flow) of a feature instance at which feature interaction resolution has to be performed (Fig.13). These points are before an event will be assigned to an interested feature or feature instance. The events can be trigger events (change of connection state) or monitor events (change of device status) and feature activation messages (see Fig. 3). The trigger and monitor events are requested by the feature instances and generated by the basic switch.

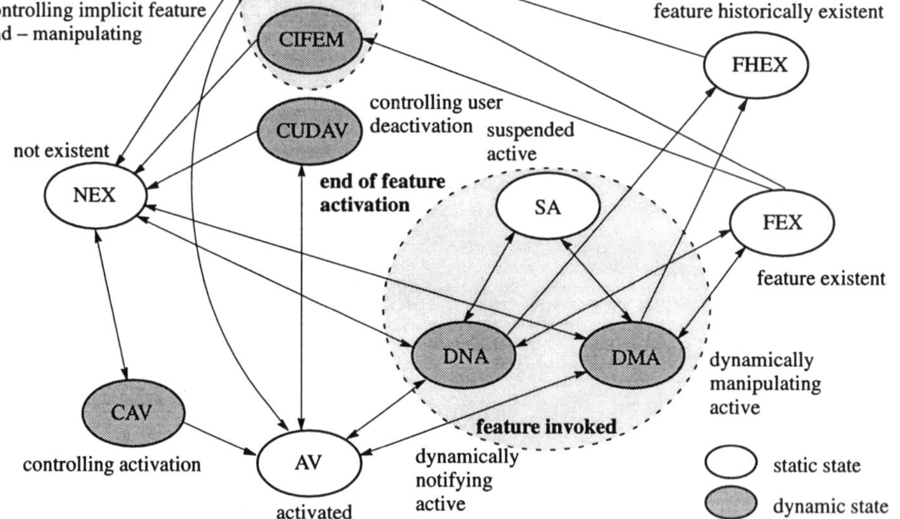

Fig.13: Generic feature state machine (GFSM)

The structure of the GFSM is defined such that the states are characterized using three different criterion:

- Dynamic – static states
- Functional: activation, deactivation, invocation, execution, termination, history
- Manipulation of call model instances within the dynamic state

First, there are dynamic states and static states. In a static state the feature is waiting for an event to continue its control flow, in a dynamic state it is active and executes its control flow. The dynamic states are coupled with the execution of feature control blocks. The whole feature control flow for one feature is divided in a message–oriented way in several feature control blocks. These feature control blocks provide all control activities, which are required when an event for the feature occurs. When an event is received by a feature instance, the corresponding feature control block is invoked. The effect is a state transition from a static state to a dynamic state. The result of the executed feature control block determines the next state of the GFSM. If the required constraints (eg. connection states, device status, feature activation parameters) for a successful execution of a feature control block are not fulfilled, the previously assumed state transition (static to dynamic state) with its corresponding control activities has not been performed. The result is, that there are several possible state transitions from a dynamic state to a static state depending on the result of the feature control block execution. Secondly, the static states are functionally structured (activation, deactivation, invocation, execution, termination) [10]. Additionally there are static states required, which represent a successfully performed feature control flow (eg. a forwarded call). Some features with special characteristics have an effect on the interaction behavior with other features or other instances of the same feature after its control flows have been finished. For example, the handling of a forwarded call will differ from an ordinary 2–party call. To express this behavior, the static state FHEX represents a successfully performed feature instance without return to a required control flow and a second static state FEX enables the return to a feature control flow to invoke an existing subfeature (eg. *Call Alternating* in a *Consultation Call* situation). The third distinguished states are the dynamic states, which are critical or uncritical states. In a critical state the corresponding feature control blocks executes functions, which manipulate the call model instances of the basic switch, in contrast to the uncritical states. In latter states no write access functions to the call model instances are performed. This distinction of states values the primitives of the feature control flow which can be executed if the event will be assigned to the feature instance by the FIM.

The GFSM enables the FIM to determine the way of using the event through the feature instance. The events will be classified by the examination of the possible state transition. The FIM does not take into account any busy or error situations, when it determines the possible next state. It assumes a straight forward behavior of the feature instance. Providing these informations, the FIM is able to transfer events to the features, if they use it uncritically. In addition the FIM has to know if the event can activate or invoke a feature. The possibilities of influencing the specified feature control flow depends on the actual state of the feature instance.

The GFSM contains all possible states and transitions, which can be passed by a feature instance. It describes all possible features, those with explicit feature activation (eg. *Call Forwarding Unconditional* CFU) and those with coincident activation and invocation (eg. *Consultation Call*).

The transitions through the generic feature state machine are described at the example feature *Call Forwarding Unconditional* (CFU). This example encompasses the activation, the invocation, the release of a forwarded call and the deactivation. T(M)Ack stands for the straight forward execution end of the corresponding feature control block (which processes the message M). A feature control block provides all required control flows when a specific event or message for the feature occurs.

Activation: The device requests the feature CFU (feature request FReq)

      FReq(CFU)   NEX $\rightarrow$ CAV
      $T(FReq_{CFU})$Ack   CAV $\rightarrow$ AV

Invocation: The event E (incoming call to CFU source) occurs

      E   AV $\rightarrow$ DMA
      T(E)Ack   DMA $\rightarrow$ FHEX (represents the existing forwarded call)

Release of the forwarded call: End of the dynamic feature state machine instance

      Release   FHEX $\rightarrow$ CIFEN
      T(Onhook)Ack   CIFEN $\rightarrow$ AV

Deactivation: The device clears the activation of CFU (feature cancel FCanc), end of the
                 feature instance

      FCanc   AV $\rightarrow$ CUDAV
      T(FCanc)Ack   CUDAV $\rightarrow$ NEX

There can exist several calls in the system, which have been forwarded by the same device (with activated CFU). This requires the creation of multiple instances of the same feature state machine for one feature instance (the feature instance is created when the feature is activated), when a call is forwarded by the device. After the release of the forwarded call, the instance of the dynamic feature state machine will be cleared and the state of the feature corresponds to the static state *activated* (AV). When the feature is deactivated, the dynamic instances of the state machines still exist. These instances will be cleared by the release of the forwarded call.

The GFSM is used to resolve event interactions. When the FIM receives an event, which could be used by multiple features in different ways, it determines all related state transitions of the features. For all uncritical transitions (no manipulation of the call model), the FIM distributes the event to all these features, which use this event in an uncritical way. In addition the GFSM provides informations for the former presented feature role mechanism (eg. can the feature be invoked, when the event is given to it).

5. Event Interactions

Event interactions occur when multiple features are interested in the same event. The following conflicts of event distribution are possible:

- Initiating: Multiple features could be invoked by using the event.
- Subsequent: The control flow of multiple features could be continued or finished by using the event.
- Initiating – subsequent: One feature could be invoked, the other could be continued or finished.

It is assumed that the feature identifier is unambiguous so that it is not possible that two different features could use the same feature activation message. These interactions are resolved using functional signalling [3].

The following generic processing rules are suggested to resolve these interactions. Some of these rules have been proposed for IN [1] [8] [9].

- Ending *before* subsequent *before* initiating (based on GFSM).
- Priority for features with the same initiating event.
- Time of the event request determines the precedence, when both features can use the event as subsequent or ending.
- There can only be one feature instance simultaneously which manipulates the call model when using the event.
- The first manipulation of the call model finishes the event distribution (based on GFSM)

If multiple features are able to use the same event and perform call model manipulations, these features access with high probability the same call model instance. The effects on the feature interactions can not be controlled, when both features are executed simultaneously. Hence it follows, that the constraint is introduced: only one manipulating feature can exist simultaneously. These critical event interactions occur with less probability if the presented mechanism for the resolution of functional interactions is applied and all elements of the feature role matrix are specified correctly. Both features want to manipulate the call model. This means they cannot be compatible, so that the two feature roles cannot coexist for one device.

Figure 14 shows an example event interaction. An ARBS call is created to a device with activated *Call Forwarding on No Reply* (CFNR), device A is alerting. CFNR is invoked by this event and a timer is started. The ARBS instance is waiting for the event, that A accepts the incoming ARBS call, to add the connection to the *ARBSPartner* B. The same event can be used by the CFNR to cancel the forwarding and to clear the timer. Applying the suggested event distribution rules, the following interaction behavior between these two features can be achieved. If B accepts the incoming call, the event is given first to CFNR (ending state transition without any manipulations) and after this it is given to ARBS (continuing state transition with call model manipulations).

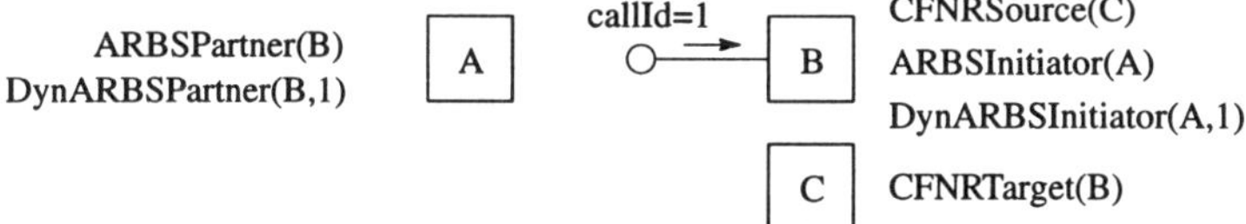

Fig.14: Event interaction *Automatic Recall on Busy Subscriber* (ARBS) and *Call Forwarding No Reply* (CFNR)

6. Combined Algorithm to Resolve Feature Interactions

The resulting feature interaction (FI) algorithm is a combination of the three methods presented, the feature role mechanism, the generic feature state machine (GFSM) and the generic event processing rules. The error interactions will be resolved by the described actions, which are included in the feature role mechanism.

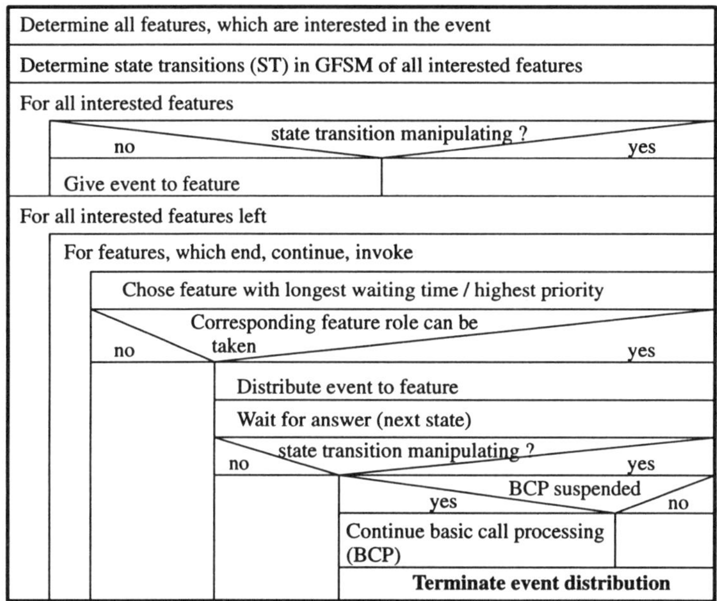

Fig.15: Event distribution

The treatment of an event message (algorithm for event distribution) is as follows (Fig. 15). First, all features are determined which are interested in the occurred event. To achieve this, the FIM has to search the event tables of the database for all feature instances, which have requested the event. For all these features the corresponding state transition is determined by examining the actual state, the event and the specified feature state machines. Then the event is distributed to all feature instances, which use this event uncritically (GFSM state transitions: NEX, AV, SA, FEX → DNA or FEX, FHEX → CIFEN). The FIM processes all remaining features in the order *ending* before *subsequent* before *initiating* state transitions. Before the event will be given to a feature instance the FIM checks if a possible corresponding feature role exists and if this role can be taken (eg. incoming call for *CFUSource* requires check of *DynCFUSource*). If incompatibility is detected the feature is discarded of the set of interested features. In the ending and subsequent case, the FIM transmits the event to the feature instance with the longest waiting time. In the initiating case, it selects the feature with the highest priority. When the event is given to a feature instance, the FIM performs the transition to the next state (dynamic state) for this feature instance. Then the feature instance processes its control flows. After this, the feature instance informs the FIM about the processed actions. This message defines the next state of the feature instance. Additionally it informs the FIM about the cause, if the assumed straight forward state transition and the corresponding control activities have not been executed successfully. If the state transition was uncritical (no manipulations of the call model instances) or the return message indicates that no manipulations have been performed, the next interested feature can be processed. The event distribution terminates, when the first of the interested features has manipulated the call model.

Only one feature of the set of interested features is allowed to manipulate the call model, so that the event distribution ends, when one of the interested features uses the event in this way.

The feature role mechanism resolves the functional interactions, when a feature activation is requested. The feature identifier is unambiguous, so that this mechanism is sufficient for handling these feature interactions.

All features explicitly request the required feature roles for the devices, which are affected by the feature control. This ensures that only compatible features are able to coexist.

The feature interaction algorithm has achieved the results shown in figure 16. The algorithm does not take into account the interactions between multiple features (more than two) which require a specific action. The interaction between more than two features are resolved by checking the corresponding feature roles in pairs. A number of 11 basic features (appendix) has been analyzed. All event interactions and error interactions have been resolved. Most of the predefined functional interactions can be achieved with the feature role method.

Feature Benchmark		
Number of realized features	11	
Functional Interactions		
Specified functional interactions	80	
Resolved functional interactions	76	**95%**
Event Interactions		
Number of detected event interactions	13	
Number of resolved event interactions	13	**100%**
Error Interactions		
Number of detected error situations	21	
Number of avoided error situations	21	**100%**

Fig.16: Results

There are some cases of a desired specified feature interaction behavior, which cannot be achieved by the feature role mechanism. These cases are detected within the phase of defining the elements of the feature role matrix. To handle most of these interaction cases there are several possibilities: to extend the role parameters, to choose another point of resolution (activation of feature f_1, f_2 or invocation of f_1, f_2) or to provide a special treatment routine.

7. Realization of the Feature Interaction Mechanisms in the Environment of a Prototype Architecture

7.1. Feature Role Mechanism

The required feature roles will be requested by the features for all affected devices. Figure 17 shows the schematic control flow when the feature *Call Forwarding Unconditional* (CFU) is requested by a device A. Before the activation message is sent to the feature, the FIM checks, if the activation device can take the corresponding role (*CFUSource*). This reduces a consultation of the feature, when the role cannot be taken by the device. This check can only be done for the activator, because the feature control flow will collect the activation parameters after it has been invoked. The activation message is given to the corresponding feature program, a new instance is created and the message is handled by the corresponding control flow for the feature activation. The feature collects the device identification of the target of the CFU from the device. Then the feature instance requests the feature role *CFUSource* for device A. The role request is checked by the FIM and if the role can be taken, the FIM performs the actions included in the role matrix and acknowledges the successful taking of the feature role to the feature program. After this the feature role *CFUTarget* is requested. If the two roles are taken successfully the feature continues its control flow, activates the initiating trigger point and performs all required control activities. If one of these roles cannot be taken, the activation procedure will be terminated, the already performed actions will be rolled back and the feature activation will be rejected to the user.

When the initiating event for the activated CFU occurs, the dynamic feature roles will be requested. Before these roles can be requested, all required feature role parameters have to be determined.

The release of feature roles can be done explicitly by a feature control flow and implicitly by the FIM. When deactivating a feature the static roles are released by the corresponding feature control flow. When a forwarded call is released by the user the dynamic roles will be released by the FIM. For this, the FIM activates specific triggers at the invocation of the activated CFU. These triggers are marked as FIM triggers, so that their occurrence does not lead to the invocation of the general feature interaction algorithm.

It is desired, that the role mechanism will be hidden to the feature control flows. That cannot be done, because of the required role parameters and the collection of activation parameters by the feature control flow. Additionally, there are constraints for the activation and invocation of features. These have to be checked by the features before the required feature roles can be requested.

7.2. Generic Feature State Machine

When the FIM receives an activation message for a feature, it creates a feature state machine for this requested feature, if the FIM logic allows a consultation of the feature program. The state transition to the corresponding dynamic state is performed. The required feature control flows are executed – this encompasses the feature role requests – and the feature informs the FIM about the status of the executed control flows by using standardized return values. The possible return values are

- normal execution end
- feature roles rejected (features functional incompatible)
- parameter error
- end of feature instance
- error in execution

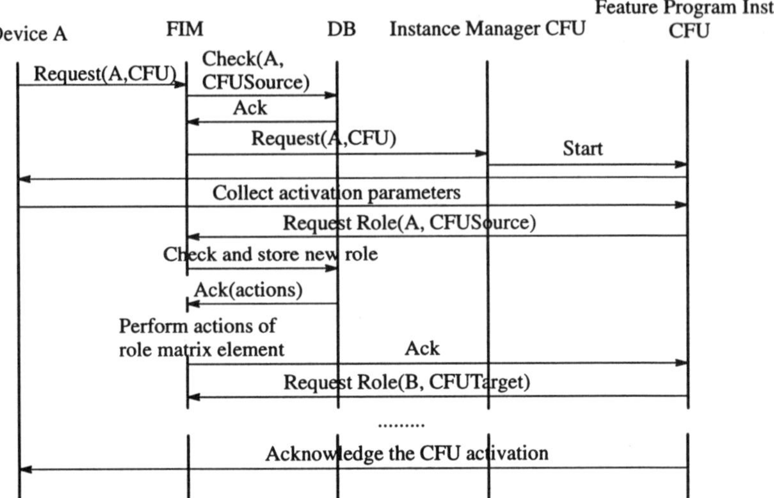

Fig.17: Request of feature roles, example activation of
Call Forwarding Unconditional (CFU)

- error in feature end

This return message leads to the required state transition to a static state, in which the feature instance waits for the next event. Figure 18 shows the management of the feature state machine at the activation and the invocation of the feature *Call Forwarding Unconditional* (CFU). The required state transitions have been explained in section 4. The state transition, when a forwarded call is released, is performed automatically by the FIM. For this it uses the same FIM triggers as described in the former section for the release of dynamic feature roles.

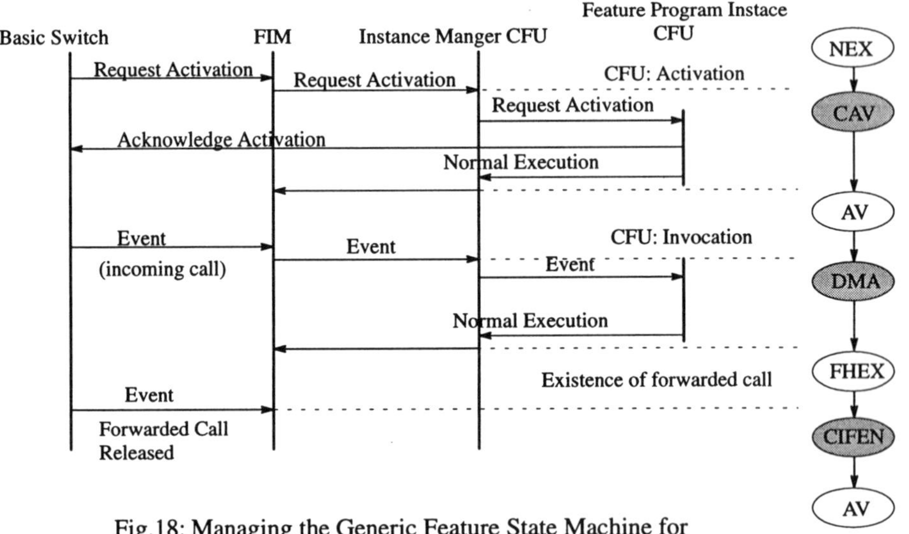

Fig.18: Managing the Generic Feature State Machine for
Call Forwarding Unconditional (CFU)

7.3. Effect of the Feature Interaction Algorithm on the Feature Logic

The introduction of the presented feature interaction resolution techniques have an effect on the structure of the feature programs.

The feature roles are explicitly requested by the feature control flows. The release of roles can be performed explicitly or implicitly. This means, that the release is visible in the feature control flow or not. Within the feature logic the control flow can only be continued, if the feature role request is successfully acknowledged by the FIM and the corresponding actions within the role matrix have been processed successfully. The consequences for the feature control flows are, that all feature roles have to be requested before any irreversible control action (manipulation of call model objects) is performed by the feature instance. In addition, the required feature role parameters have to be known, when a feature role is requested.

The information for the generic feature state machine, which initiates the state transition from a dynamic to a static state have to be provided by the feature instances after its control activities have been executed.

The resulting structure of a feature control block, which is responsible for the processing of one feature event is shown in figure 19. Each of the activities within a feature control block can result in the termination of the feature control flow. The corresponding return value will be notified to the FIM, which initiates the required transition in the feature state machine.

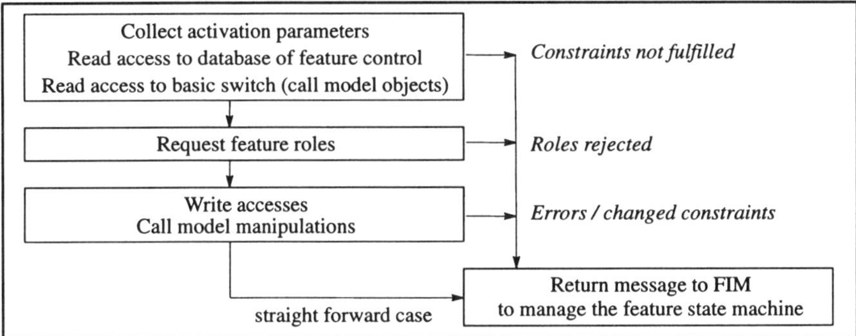

Fig.19: Effect of feature interaction algorithm on feature control blocks

7.4. Prototype

The architecture with separated call and feature control has been implemented in prototype form. The SDL (Specification and Description Language [11]) tool SDT from TeleLOGIC [12] has been used to achieve an executable application.

The prototype (Fig.20) consists of the basic switch, the feature control node and a configurable number of graphical user interfaces for the devices. The basic switch and the feature control are specified in SDL. The feature interaction algorithm requires complex access capabilities to its data structures (eg. feature roles, feature states, activation parameters). To manage the data the mechanisms provided by SDL are not sufficient. Therefore the database is implemented using the relational database management system INFORMIX SQL [13]. The graphical device interfaces are implemented in 'C' with a window system library. All these components communicate via a tool, which is provided by the tool SDT. This enables SDL systems to exchange signals automatically corresponding to the specification of their environment interfaces.

The graphical device interface enables the user to set up calls, to release calls and to activate and deactivate features. Additionally, the interface provides output mechanisms to view the actual call topology from the view of the device and to view the actual taken feature roles.

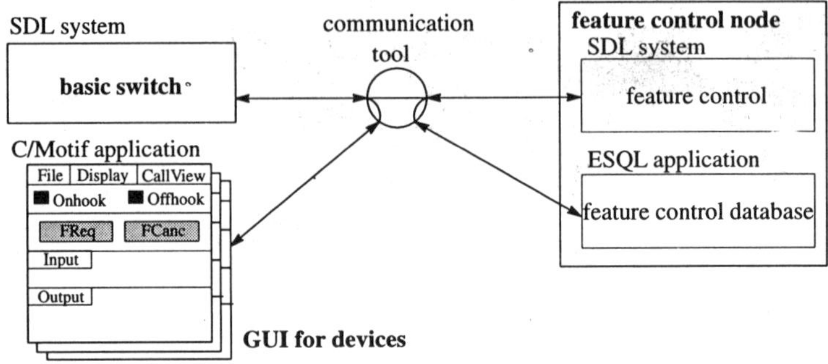

Fig.20: Architecture of the prototype

8. Conclusions

In this paper a new approach to resolve most of the feature interactions in an architecture with separated call control and feature control has been presented. The feature interactions are classified in 3 distinguished kinds, the functional, the event and the error interactions. For each class a mechanism has been presented. The specified feature interaction behavior has been obtained by the introduction of the feature role mechanism. To prevent error situations because of architectural and modelling constraints additional actions can be included in the feature role matrix. The generic feature state machine (GFSM) and the event distribution rules resolve conflicts, when multiple features can be triggered by the same event. The realization of the feature interaction algorithm and its effect on the feature programs has been described.

The feature interaction algorithm presented has been developed and tested in the environment of a prototype of an architecture with separated call and feature control. As a feature benchmark, 11 common features (see appendix) have been developed and their functional interactions have been specified before the feature interaction algorithm has been developed. Using this set of functional interactions the feature interaction algorithm has been verified and validated.

Acknowledgements

This work has been funded by Siemens AG, Private Communication Systems, Munich. Many fruitful discussions with E. Wildgrube and his colleagues as well as with Prof. J. Eberspächer from Technical University of Munich are gratefully acknowledged.

References

[1]   ITU–T Q.1200 Series, Intelligent Network

[2]   N. Fritsche, Basic Call Processing Architecture for Flexible Control of Supplementary Services, Proceedings of International Switching Symposium ISS, Vol. 2, pp. 137–141, April 1995

[3]   E.J. Cameron, N.D. Griffeth, Y.–J. Lin, M.E. Nilson, K. Schnure, H. Velthuijsen, A Feature Interaction Benchmark for IN and Beyond, Proceedings of Second International Workshop on Feature Interactions in Telecommunications Software Systems, IOS Press Amsterdam, pp. 1–23, May 1994

[4]   E.J. Cameron, N.D. Griffeth, Y.–J. Lin, H. Velthuijsen, Definition of Services, Features and Feature Interactions, Bellcore Memorandum for Discussion, International Workshop on Feature Interactions in Telecommunications Systems, December 1992

[5]   T.F. Bowen, F.S. Dworak, C.H. Chow, N. Griffeth, G.E. Herman, Y.–J. Lin, The Feature Interaction Problem in Telecommunications Systems, Proceedings of 7th International Conference on Software–Engineering for Telecommunication Switching Systems, pp. 59–62, July 1989

[6]   N.D. Griffeth, H. Velthuijsen, The Negotiating Agent Model For Rapid Feature Development, Proceedings of 8th International Conference on Software Engineering for Telecommunication Systems and Services, 1992

[7]   L. Schessel, Administrable Feature Interaction Concept, Proceedings of International Switching Symposium ISS, Vol. 2, pp. 122–126, October 1992

[8]   Bellcore, Multi–Vendor Interactions MVI, Compendium of 1989 Technical Results, Special Report SR–TSY–001629, Issue 1, March 1990 – Part D, pp. 54–61

[9]   Bellcore, Advanced Intelligent Network, Release 1 Network and Operations Plan, Special Report SR–NPL–001623 Issue 1, June 1990

[10] ETSI, Network Aspects, Intelligent Networks, Service and Service Feature Interaction, Service Creation, Service Management and Service Execution Aspects, ETSI draft ETR NA611–01, Version 0.8, February 1995

[11] CCITT, Functional Specification and Description Language (SDL), Blue Book Recommendation Z.100, 1988

[12] TeleLOGIC Malmö AB, SDT (SDL Design Tool) Version 2.3, 1993

[13] INFORMIX, Guide to SQL, Tutorial, 1991

[14] CCITT, Message Sequence Charts (MSC), Recommendation Z.120, 1992

APPENDIX

A Feature Benchmark

CFU	Call Forwarding Unconditional
CFBS	Call Forwarding on Busy
CFNR	Call Forwarding on No Reply
TCS	Terminating Call Screening
ARBS	Automatic Recall on Busy Subscriber
CONF	Conference, Add Party to Conference
CC	Consultation Call (includes Call Hold)
CA	Call Alternate
CT	Call Transfer
CO	Call Offer, Take Call Offer
AB	Abbreviate Dialing

Feature Interactions in Telecommunications Systems III
K.E. Cheng and T. Ohta (Eds.)
IOS Press, 1995

Minimizing Feature Interactions:
An Architecture and Processing Model Approach

Israel Zibman, Carl Woolf, Peter O'Reilly
GTE Laboratories, Inc., 40 Sylvan Road, Waltham, MA 02254, USA

Larry Strickland, David Willis, John Visser
BNR, PO Box 3511 Stn C, Ottawa, Ontario K1Y 4H7, Canada

In this paper we propose an approach that uses an architecture in conjunction with a processing model to minimize the problem of feature interaction. Our approach recognizes that currently many assumptions about service operations and system capabilities are implicitly built into the service software architecture. We claim that many non-trivial feature interactions arise when services or technologies with new capabilities, based on conflicting assumptions, are introduced. We propose an architecture based on software agents and separation of concerns. This architecture effectively removes many common assumptions about roles and technology from the service software architecture, thus reducing the need to provide *ad hoc* solutions to particular feature interactions. We combine this architecture with a processing model that ensures service independent management of feature interactions. We demonstrate the power of such an approach by showing how it deals with examples of interactions among existing services, and interactions introduced by Universal Personal Telecommunications (UPT).

1. Introduction

Feature interaction (FI), in the context of telecommunications, is the phenomenon where a user (end-user or system operator) observes services failing to perform as expected, and where the failures are due to the presence of other services (or multiple instances of a single service) in the network. Addressing FI caused by new services needing to inter-work with existing services causes significant expense and delivery delay for new services. Often the only solution available is to modify the existing service software or even the service specifications to enable them to interwork with the new services [1][2].

Causes of FI have been extensively examined in the literature [2][3]. These include signal ambiguity, lack of information for handling conflicts, and typical problems of distributed systems. In looking for a root cause of the problem it appears that unintended FI occurs when previously held assumptions are violated by new services. New services have new needs that were not anticipated when existing services and technology were deployed.

These existing services and technology thus made many assumptions about the roles they needed to use or play. For instance the role of the directory number (DN) in older services is assumed to be both the billing number for originating calls and the terminal location. New mobility services will need to separate the roles of billing number and terminal location, thus violating a previously held assumption. Likewise freephone services no longer associate the bill payer with the call originator. These assumptions are often built into the hardware and software, causing FI to occur when new services are introduced that use a different set of assumptions.

There are various types of methods for addressing FI. In terms of the taxonomy of methods proposed by Bouma and Velthuijsen [4], our approach is of the 'avoidance' type. We propose an approach that consists of two components. Like some other researchers [5][6], we suggest an architectural framework in which the complexity of distributed telecommunications systems is reduced by identifying distinct components of the systems and by keeping their concerns and responsibilities separate and well-defined. We also propose the use of a processing model, within the architecture, for managing interactions in a service independent manner

Section 2 of this paper discusses sources of interactions in the form of violation of assumptions and role confusion. An architecture, based on agents and separation of concerns, is introduced in Section 3 as an approach to managing FI. The processing model for service-independent interaction management is discussed in Section 4. In order to demonstrate the power of the Agent Architecture, Section 5 applies the Architecture to features with known interactions, and new features (such as those introduced by Universal Personal Telecommunications - UPT [7]). A discussion of further work is provided in Section 6 before we conclude in Section 7.

2. Violation of Assumptions and Role Confusion: Sources of Interactions

Violation of assumptions is a major source of interactions. These interactions typically arise when new features and network capabilities which violate pre-existing technology assumptions are introduced. The usual response is to complicate definitions and implementation for existing services to cope with each new change in the environment. As a result, the system becomes less testable and less resilient to future changes.

As an example, consider the call-waiting service. The basic idea of call waiting is to alert the user that a new call has arrived and to provide a mechanism for the user to respond to the call. Ideally, the service logic should be that simple. In practice, the service logic embeds large numbers of assumptions about limitations on when an alert may be delivered (only when a call is 'stable'), what kind of alert should be delivered (particular 'beep'), how many calls may be in the 'waiting' state (one), and how to respond to a call (hook-flash). Although these constraints are relevant in a POTS environment, they do not make sense to systems with ISDN, sophisticated terminals, or more capable bridges. The constraints also lead to problems if the user has other services which use the same signals for different purposes. Such constraints should not be part of a service definition and should not be coded into implementations. Instead, they should be dealt with by allowing a richer set of interactions between network resources and service logic. Our proposed architecture supports these interactions.

The concept of roles is also useful to understanding the types of interactions that can arise from violation of assumptions [3]. Many roles are played by various entities in the current telecommunications system. Users may fill roles of owner, subscriber, bill payer, caller, called party, call coordinator and others. Services fill roles of busy treatment function, call termination blocker, call redirector, and many others, replacing the default behavior of ordinary call processing. Lack of separation between the various roles that these entities play is a significant source of FI in current systems. Role confusion is manifested in the following ways:

- Billing might go to the wrong party, for example, when a service hard-codes an assumption that the owner of the phone is the service user.

- Calls might be delivered when undesired, blocked inappropriately, or delivered to the wrong party because the system could not properly determine the destination of the call.

Role confusion contributes to signaling ambiguities in part because service logic often appears to assume that a single service is the only user of a particular signal or feature. An example of this is the hard-coding of a specific alerting pattern into the logic for a distinctive calling service. If the service really intends to inform the subscriber that a special event has occurred it must be sure that the alerting signal is unique. Similarly, service logic often assumes that no other service will use the hook-flash. Services sometimes assume that the only relevant notion of 'busy' is the event reported by the switch, ignoring the possibility that a set of services could create a logical hunt group, in which case 'busy' should mean that all terminals in that group are in use.

3. An Architecture based on Separation of Concerns

Previous work on the sources and causes of FI [8] and on analysis of telecommunications feature descriptions and broadband service requirements has shown that specific compartmentalizations, or separations, of concerns within a telecommunications system will help to reduce or eliminate many cases of FI. Specifically, separating:

- users from terminals

- calls from connections

- user sessions from services, and

- services and user sessions, from management of network/resources

presents significant opportunities to reduce interactions which may arise in a telecommunications system. Separation also implies breaking the system down into components each of which can be assigned certain roles.

We propose an Agent Architecture which provides a framework and specification support for separation of concerns. Separation is achieved by supporting separate logic for

- programs that enforce user profiles and preferences,

- programs that control resources such as terminals, switches, bridges, interactive voice (or video) response units, and

- programs that carry out the basic logic of services.

This separation of concerns is achieved through the use of software agents representing each of the types and instances of entities - users, resources, and services - in our Agent Architecture, each with its own explicit concerns. The goal of the Architecture is to greatly reduce the number of assumptions which are embodied directly in the Architecture and thereby to increase the probability that new services will work properly with existing services. Our Agent Architecture is depicted in Figure 1.

The Architecture consists of a set of agents that work together to support telephony services for end users. The Architecture builds upon one originally proposed in [8]. Each agent has particular roles in the network, and behaves accordingly. There are currently four types of agents in the Architecture: User Agents, Connection Agents, Resource Agents, and Service Agents. The basic characteristics of each type of agent, along with their characteristics which minimize FI, are described below. Attached to each agent is data and program logic which enables the agent to play its roles.

User and terminal separation is achieved through the creation of distinct User and Resource (Terminal) Agents; call and connection separation is achieved through the management of calls by User Agents (i.e. through sessions) and connections by independent Connection Agents which coordinate multiple Resource Agents; user session control and service separation is also explicit in the selected agents; and identification of Resource Agents allows for separation of resource management from session control and services.

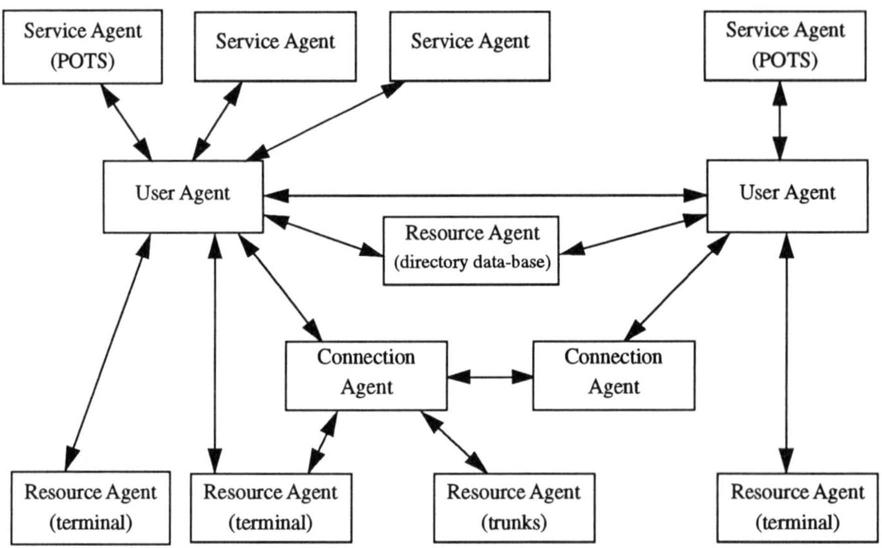

Figure 1: Agent Architecture

3.1. User Agent

The User Agent represents the user and supports communication session control on the user's behalf in the network. It is possible that a particular end user might have several User Agents, representing different aspects of the user's life that communications services might address. For example, the user might have a home persona and a work persona, each represented by a different User Agent. It is also possible for one User Agent to represent multiple users, for instance a family which wishes to share the same User Agent.

One major responsibility of the User Agent is to manage any calls that the user is involved in, i.e. to manage the user's communication sessions. Communications between users are initially established and coordinated by the User Agents of the parties involved. To establish an outgoing call a User Agent would first communicate to other User Agents the desire for such a call. Once the User Agents have agreed on whether a call is feasible and have agreed on destination information for call completion, the originating User Agent will then request that the Connection Agent coordinate the bearer capabilities required. This approach results in call/connection separation and negotiated connection setup. The independence of the negotiation phase from the connection setup simplifies the resolution of interactions which can arise from restriction (e.g. screening) features.

The user's preferences and profile are expressed in the choice of services subscribed, subscriber-dependent data accessed by those services, and the User Agent rules regarding dispatching of messages and events to Service Agents. The User Agent is able to fill the roles of billing party, called party, terminal owner, etc. through the use of Service Agents which are designed to respond to events concerning these roles.

Section 4 proposes a processing model, implemented in the User Agent, for controlling services and message passing in order to allow FI to be resolved in a service independent fashion.

3.2. Service Agent

Each Service Agent represents the processing needed to carry out the intentions of a service. Each subscribed Service Agent indicates to the User Agent which events the Service Agent is interested in. A Service Agent responds to events by providing information to the User Agent on how to execute the service. As part of the response to an event the Service Agent may provide new indications of events of interest to the User Agent.

It is essential that Service Agents be designed so that they avoid hard-coded assignments of actors to roles, in particular the hard-coding of specific signals. Signals should be expressed in terms of logical names (e.g. "teenage caller alert", not "distinctive ringing pattern 2"). Service Agents must be designed to respond to events indicating service activation or new terminal assignment. In response to these events the Service Agent must ask the User Agent to query the Terminal Agent for a suitable signal. Service Agents need to be prepared for the event that not enough unique signals are available. They must also provide mechanisms for informing the human user about which signals will be used.

One particular Service Agent is the POTS agent which is designed to manage basic calls originating or terminating at a terminal registered to a user. The details of a POTS Service Agent are presented in a later section. The event interest indications for the POTS agent allow other Service Agents to pre-empt the POTS Agent. By having POTS represented as a service like any other we avoid the architectural disadvantages of needing to treat 'special services' and POTS differently [8].

3.3. Connection Agent

The Connection Agent represents the network provider in the establishment of bearer channels. The Connection Agent has algorithms for establishing connections between different physical locations. The locations are identified to the Connection Agent in requests from a User Agent or another Connection Agent. The request can include desired constraints such as acceptable cost, minimum bit rate, optimal bit rate, and maximum delay, as well as the two physical locations between which a bearer channel is to be established and any extra functionality required such as conference bridges.

The Connection Agent in turn makes its own requests to Resource Agents to establish a bearer channel. If it does not have access to all the resources required it might send a request to another Connection Agent, to manage another part of the connection.

3.4. Resource Agent

Resource Agents have the role of managing functionality contained in resources. They keep track of available functionality, functionality in use, functionality reserved for a particular user or other agents, and whether User Agents need to be informed when some particular functionality is accessed. A special case of a Resource Agent is a Terminal Agent which represents end-terminals. A Terminal Agent can accept requests from both Connection Agents and User Agents.

An important function of a Resource Agent is the reduction of signal ambiguity. At subscription time or service set-up time, a Service Agent will ask the User Agent to send a message to a Resource Agent requesting a signal satisfying particular characteristics. For example, a Call-Waiting Service Agent might ask a Terminal Agent for an alerting pattern applicable while a call is in progress. The Terminal Agent might respond with a unique signal or it might indicate that no unique signals are available. In the latter case it will be prepared to respond to queries about how appropriate signals are currently assigned. Service Agents could then deal with how or whether a signal could be shared and how the user should be notified.

Resource Agents also keep track of which User Agents are using busy resources, and may have rules for how to respond, or what User Agent should be consulted, if a different User Agent tries to control a busy or reserved resource.

Resource Agents and Connection Agents can require billing authorizations for the use of their resources. Service Agents which require use of these resources may need to obtain authorization from other parties (represented by User Agents) if billing is not to be paid by the requesting User Agent.

4. A Processing Model for the Architecture

In order to remain service independent in the design of the architecture it is desirable that the

User Agent be able to make decisions which resolve interactions independently of the specific services which have been subscribed. To achieve this goal, we are proposing a processing model which allows the User Agent to make decisions based entirely on the intelligence contained in the Service Agents.

4.1. Role of the User Agent

We propose that the User Agent embody very little information and function primarily as a gateway and dispatcher (i.e. coordinator) for messages between outside agents (Connection Agents, Resource Agents and other User Agents) and the Service Agents supporting the services subscribed to by the user. The User Agent uses a set of dispatch rules to determine when a given Service Agent should receive a message, and when a message should be sent to an outside agent. Since the User Agent is expected to remain deployed without modification as new services and technologies are deployed and, perhaps, older services and technologies retired, the language for expressing these rules cannot depend on any service specific concepts.

4.1.1. Message Handling

The User Agent maintains a list of 'incoming messages' which have arrived from outside agents and have not been responded to and a list of 'outgoing messages' which are being readied for transmission to outside agents. Each message contains parameters which may be referenced in some service independent fashion such as ASN. We assume that there exist function calls which can ascertain the existence of particular parameters in a message and can examine the values of parameters. We also assume a mechanism to provide call-by-reference or result-copying, so that parameters in messages can be changed by Service Agents.

We define an event as any change to the list of incoming or outgoing messages or to the parameters within such messages. Note that messages to the User Agent may occur other than in the context of classical call processing. One example of such a message is the notification to a User Agent that the user has added a new service. This event interests the User Agent since Service Agents representing existing subscribed services may wish to respond to such messages about new services, e.g. to aid in avoiding signal ambiguities.

4.1.2. Dispatcher

The User Agent maintains a list of dispatch rules governing processing of events. At service subscription time, a Service Agent sends a notification of its presence to the User Agent along with a list of dispatch rules indicating when that Service Agent should respond to an event. As part of a response to an event a Service Agent may place or modify a message on the outgoing message list and may also provide additional rules for when it wishes to respond to future events. These additional rules are, in effect, requesting subsequent events from the User Agent.

A general constraint about pending outgoing messages requires a User Agent to send them when and only when all message lists are stable. This stability occurs when none of the User Agent's service-supplied dispatch rules express interest in any incoming or outgoing message, and when all messages from the User Agent to its associated Service Agents have been answered. This constraint gives Service Agents the opportunity to respond to as well as influence the content of messages.

4.1.3. Dispatch and Precedence Rules

The dispatch rules used by the User Agent could take the following form:

"If there is (is not) a message in my incoming/outgoing list from/to such a type of agent and the value of a parameter is (is not) such and such, and the value of ..., and there is (is not) a message ... then notify Service Agent X and await a response".

Dispatch rules for Service Agent X are specified by Service Agent X alone to ensure that separation of concerns is maintained and that no side effects on unrelated services arise.

Attached to the rule is a statement that Service Agent X *might* respond to this event by, for example, creating a new message, setting a parameter to a non-null/null value, or another similar action involving messages. Note that the service is not claiming that it will always respond with such an action, but only that the action is one possible response. A service could even claim several, perhaps conflicting, responses to a given event. Typically, the claim indicates what will occur if the service executes its purpose. For example, a service which selects certain calls for forwarding, but allows others to continue normal processing, would claim that in response to a particular event it might set the operation-parameter of a message to FORWARD CALL. These claims constitute an expression of the intention of a service. The User Agent uses these "intentions" to resolve conflicts among services.

When the dispatch rules available to a User Agent would result in more than one service responding to an event the User Agent should log the occurrence of the ambiguity and make an arbitrary choice as to who should respond. Such ambiguities are likely to be sources of FI. To address some of these ambiguities the dispatch process has been extended to include precedence rules based on service intent. Precedence rules are framed in the language of events and do not make reference to any particular services. A precedence rule has the following form:

"If {dispatch style rule} then deliver event to Service Agents which claim they *might* set parameter #n of an outgoing message of type m to non-null/null (or create such and such types of messages) prior to delivering to agents which do not claim to set parameter #n or create such messages."

Precedence rules are provided to the User Agent by a Service Agent at subscription time or by users at any time they might wish to modify service profiles. User-provided precedence rules could give preference to call waiting delivery over forwarding on busy ("when BUSY CONDITION give preference to services claiming to set operation to OFFER CALL") or to forwarding to particular DNs or area codes ("... give preference to services claiming to set called party DN to match 'pattern description' ").

4.2. Dispatch and Precedence Rule Illustration

To illustrate the function of the dispatch and precedence rules in the processing model consider the following example:

A user has subscribed to a serial calling service which sequentially calls parties from a list. When each call terminates, or the party is determined to be unreachable, the service tries to connect to the next party in the list. The initial event for the service is the receipt of a message from a Terminal Agent where the dialed digits parameter contain a particular service code. When requesting a Connection Agent to attempt a connection, the Service Agent may give a subsequent event rule to the User Agent asking for messages from the Connection Agent indicating that a connection was unable to be established and the call cannot complete. This service intends to respond to such an event by generating a disconnect message to the Connection Agent and then attempting the next party on the list.

The user has also subscribed to a special find connection service which, in response to its initial event of an 'connection request nack' message from a Connection Agent intends to respond with a non-null list of new connection options (which include changes to parameters such as bandwidth, delay, cost, etc.) to be sent to that Connection Agent or another Connection Agent (representing another carrier).

Finally, the user has subscribed to a connection sorting service which, in response to an initial event of a non-null list of connection options in an outgoing message to a Connection Agent will sort the list, perhaps in order of cost or expected quality. It may delete some connection options it has determined are undesirable under current circumstances.

It is clear that the find connection service should pre-empt the serial calling service; perhaps it can find a way for the call to complete. In that case, the serial calling service need never learn that the call was in danger. This is achieved with a precedence rule similar to the following:

"When there is a message from a Connection Agent and the cause parameter is 'CONNECTION NACK' and there is no outgoing message to the Connection Agent with a non-null connection options list, give precedence to services which claim they might set the connection options parameter to non-null in an outgoing message to the Connection Agent."

If the 'find connection' service is successful, the conditions for the serial calling service to respond will not be met (since there is an outgoing message to a Connection Agent with a non-null routes list). The conditions for the 'connection sorting' service will be met. That service may delete all the suggested connection options and set the route list to null. In such an event, the event would pass back to the find connection (which kept its own state as to what it had already suggested for this call) which might have nothing more to add. The User Agent would then send the event to the serial calling service.

The precedence rules make sense and fulfill their function even if the user did not subscribe to one or more of the above services.

4.3. Processing Model Issues

Our User Agent processing model is designed to allow service interaction without requiring services (or an arbiter) to have specific knowledge of (other) services. Achievement of this goal is difficult.

4.3.1. Role of Service Designer

Service designers must avoid 'selfish' assumptions, such as the assumption of sole control over a signal or resource. Service designers should expect that any signal or resource will need to be shared, and services should be prepared to share their resources and respond to messages related to resource sharing.

The content and types of messages passed between Agents and used to define dispatch rules is subject to change as services are created or modified. These changes will affect the stability of the events, operations and parameters used to define the processing model dispatch rules. In particular these changes may arise due to incompleteness or insufficient granularity in the specification of the dispatch rules for the processing model. Ideally, new parameters could be added to existing messages without diminishing the ability of existing services to function in terms of the subset of older parameters. There must be agreement among service creators on the meaning of existing parameters. This agreement is necessary both to ensure that services use these parameters correctly, and to avoid a situation in which service creators add unnecessary or inappropriate new parameters to messages, due to disagreement about or faulty understanding of the existing set. This potential problem with the dispatch rules used to support the User Agent has many similarities with other attempts to solve the feature interaction problem. For instance in the Morgan system [9] an attempt was made to eliminate feature interactions by recasting (rearchitecting) the system at a high level. Although the interactions were successfully eliminated at that level they reappeared at another lower level involving data storage and methods.

4.3.2. Other Issues

Some mechanical issues with respect to the processing model's operation must be resolved. Care must be taken so that the dispatch rules do not lead to infinite looping. Counters or some other mechanisms need to be provided. A support mechanism must be provided that logs the occurrence of ambiguities at run-time so that they may be resolved later by service designers.

Finally, there is an open question as to how the processing model will deal with a new service that splits a pre-existing single role into several new roles. If a service creator realizes that such a differentiation is occurring the creator could create service agent actions which respond to the use of the old role (e.g. a billing role is used when a message contains a billing authorization). If the context is clear, the agent would modify the message to assign the correct actor. If the context is not clear, the event should be logged. The log would provide a warning to service designers/maintainers that the older service might need to be modified. A danger is that a proliferation of "role guardians" must stay active until we are sure that all services have been updated.

5. Applying the Agent Architecture to Specific Features and Their Interactions

In this section we illustrate how our architecture handles POTS calls, as well as several cases of FI. These examples are not intended to suggest specific service designs. Indeed, we expect that there are various ways these services might be designed. By considering specific ways the services might work, our examples support the general claim that the services and FI solutions under consideration are feasible.

5.1. Basic Call Scenario

Our first example shows the steps involved in establishing a basic call from user D to user B.

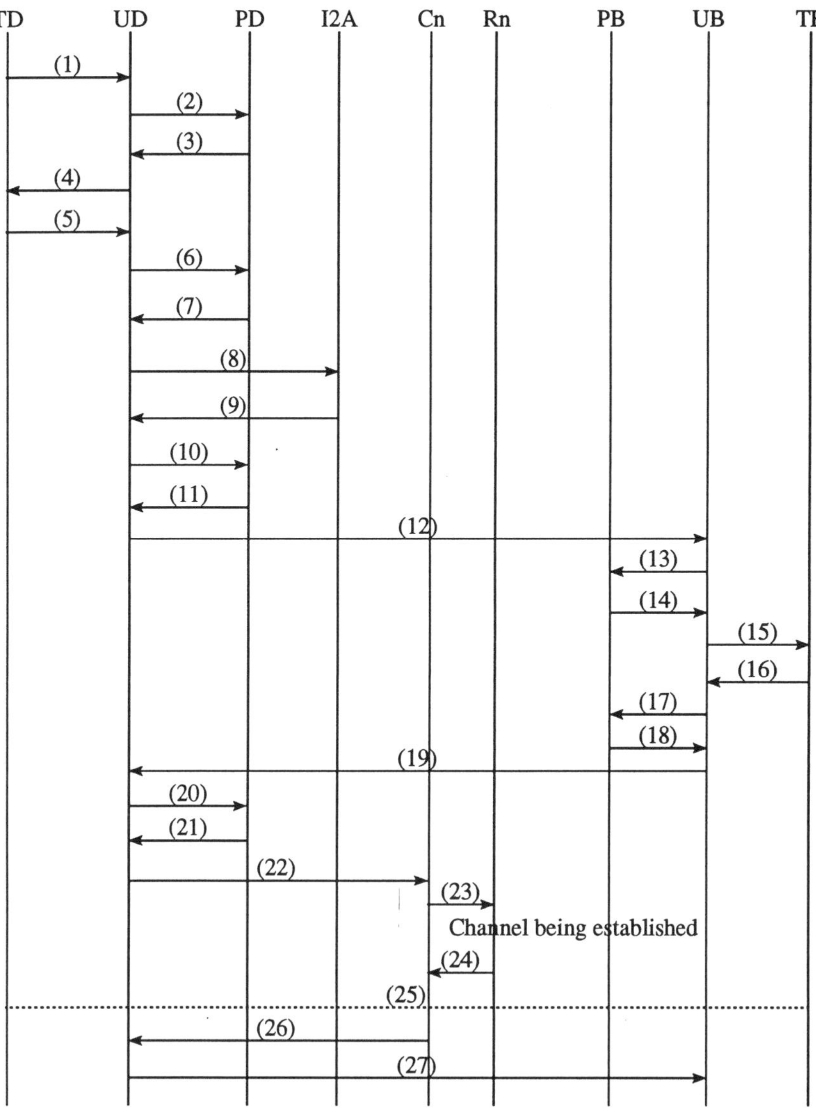

Figure 2: Basic Call Scenario

(Numbers refer to Figure 2.)

Note: For our examples we assume that only those Service Agents explicitly cited have rules expressing interest in a given set of messages. Precedence rules would be required in cases where more than one Service Agent associated with a User Agent has rules that apply to the same state of the current messages. We also assume that there are no unanswered messages from User Agents to Service Agents other than ones mentioned explicitly. Otherwise outgoing messages would need to be delayed until all Service Agents had responded.

(0) D goes off hook, which is detected by the Terminal-Resource Agent (TD) for D's cur-

rent terminal.

(1) Upon activation, TD notifies a certain User Agent, presumably the agent corresponding to the owner or administrator of the terminal. For this example, we assume D's User Agent (UD) is the relevant one. So TD sends a TerminalActive message to UD.

(2) UD has a dispatch rule, supplied by the POTS Service Agent, indicating that D's POTS Service-Agent (PD) is interested in the TerminalActive message. So UD passes the TerminalActive message to PD.

(3) PD passes back to UD a message to be sent to TD indicating readiness to accept input from the user.

(4) UD sends the message to TD.

(5) TD indicates readiness for input (e.g., by applying dial tone), and collects information entered at the terminal. The information entered in this example is 'Call B.Citizen@work.ottawa.ca'. The DN of 'B.Citizen...' (DNB) is passed to UD as the RawCalledID parameter of a ConnectRequest message to be sent to the User Agent of the called party.

(6) UD has a POTS-supplied rule expressing interest in ConnectRequest messages; so UD passes the message to PD.

(7) PD creates a name-translation query to be sent to the IdToAgentTranslation Resource Agent (I2A) in order to determine the network address of the user agent corresponding to DNB (UB). PD passes this message to UD.

(8) UD sends the message to I2A.

(9) I2A sends the requested information back to UD in a query response message.

(10) UD has a POTS-supplied rule expressing interest in the query-response; so UD passes the response to PD.

(11) PD creates a ConnectToAgent message to be sent to UB, requesting it to accept the call from UD.

(12) UD sends the connect-request message to UB.

(13) (14) UB has a rule that when there is an incoming message requesting a connection but no outgoing message to a Terminal Agent, then the incoming message should be given to B's POTS Service-Agent (PB). So UB gives the connection request to PB. Let us assume that UD has requested a 64kbit voice connection with user B. PB finds an appropriate Terminal-Resource Agent (TB) for the specified connection-type, and creates a message to TB requesting it to reserve the indicated functionality. PB passes this message to UB.

(15) UB sends the message to TB.

(16) TB reserves the requested functionality and responds affirmatively to UB, indicating a physical address to which the call should be terminated.

(17) (18) PB has supplied a rule expressing interest in the response from TB; so UB passes the response to PB. PB creates a ConnectionAllowed message to be sent to UD, agreeing to a connection and including the physical address supplied by TB. PB passes this message back to UB.

(19) UB sends the message to UD.

(20) (21) PD gets passed the ConnectionAllowed message. It creates a request to the Connection Agent (Cn - the n is used to indicate that it is one of potentially many Connection Agents) to establish a 64kbit bearer channel between the two physical locations - B's from the ConnectionAllowed message, and D's from previous interaction with TD. PD

passes the request for Cn back to UD.

(22) UD sends the request to Cn.

(23) Cn handles the request from UD by sending requests to the Resource Agents (Rn) required to establish the bearer channel, including TB, TD, and Trunk Agents, and perhaps to other Connection Agents if it does not have direct contact with all the necessary Resource Agents.

(24) The Rn establish their components of the bearer channel and notify Cn of the status.

(25) At this point an end to end bearer channel has been established.

(26) (27) Cn notifies UD, which in turn notifies UB, that the bearer channel has been established.

5.2. *UPT Delivery with Call Waiting*

We now present a slightly more interesting scenario to illustrate call processing in our architecture.

In this example, A and B are Universal Personal Telecommunications (UPT) subscribers who are currently both registered for incoming calls at a certain terminal (Tx).

B also subscribes to CallWaiting (CW), but A doesn't.

A is engaged with a call to C.

D calls B.

(Numbers refer to Figure 3.)

(0) D dials the UPT-defined DN for B (DNB). (Or, with broadband etc. in mind, 'enters the UPT-defined identifier for B'.) The call proceeds as in the basic call scenario through step (12). (13) (14) UB has a rule that when there is an incoming message requesting a connection but no outgoing message to a Terminal Agent, then the incoming message should be given to B's POTS Service-Agent (PB). So UB gives the connection request to PB. PB then creates a message to the appropriate Terminal Agent, Tx, having established the identity of the Terminal Agent when B registered for UPT service at terminal x. The message requests Tx to reserve functionality for an incoming call. PB passes this message to UB.

(15) UB sends the message.

(16) Tx refuses, indicating to UB that the terminal is busy.

(17) UB has a CW rule expressing the interest of B's CW Service Agent (CWB) in this refusal, so UB gives the message to CWB.

(18) CWB creates a message to Tx, requesting it to reserve functionality (including a CW-alert) for B's call. It passes this message to UB.

(19) UB sends the message to Tx.

(20) Tx notices that the requester, B, is not the current user, A. Tx sends a message to the current user's User Agent (UA) reporting the potential conflict between A and B over whether to allow a CW-alert. (An alternative procedure would be to consult the agent of the terminal-owner, rather than that of the current user.)

(21) (22) A's ConflictResolution Service-Agent (CRA) has supplied a rule expressing interest in, and so gets passed, the message from Tx. It creates a response to Tx, determining, let's say, to go ahead and apply the CW-alert. It passes this response back to UA.

(23) UA sends the response to Tx.

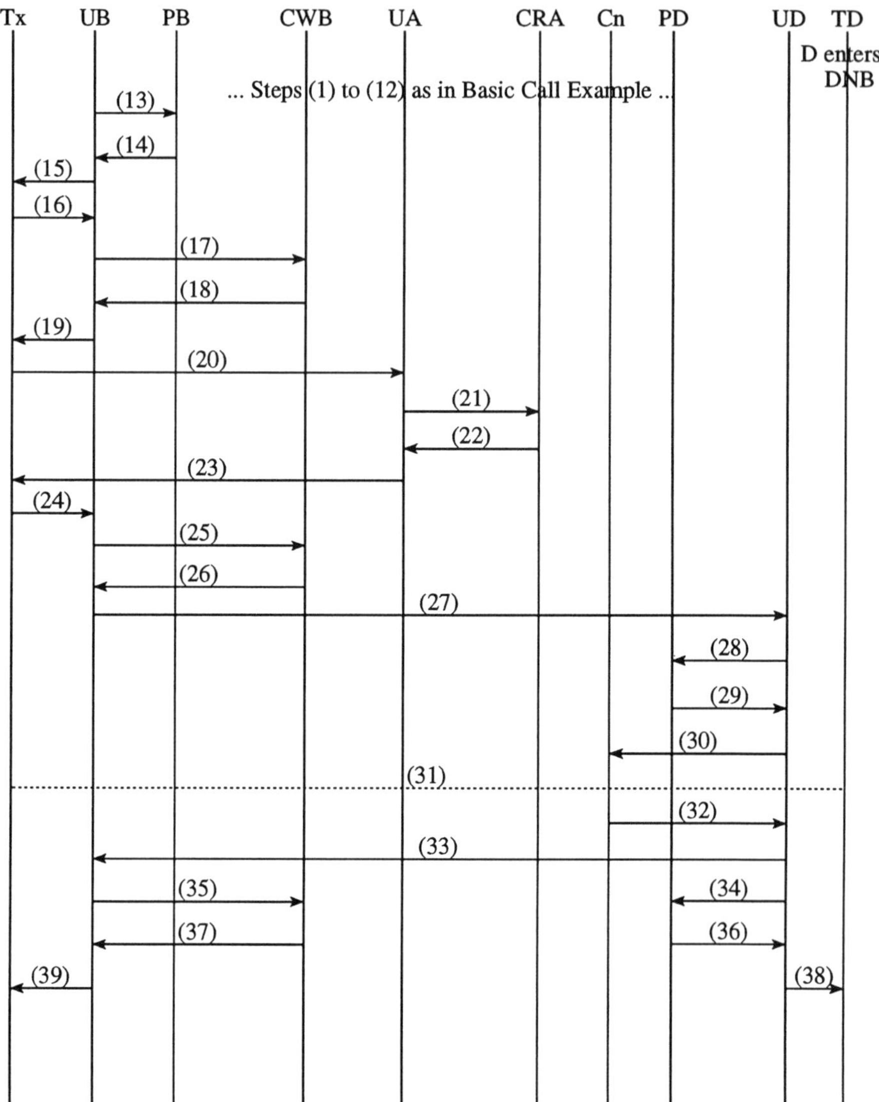

Figure 3: UPT Delivery with Call Waiting

(24) Tx reserves resources in order to alert and informs UB that it has done so, passing along its address for a possible connection.

(25) UB has a CW rule that expresses interest in messages whose parameters indicate that CW-alerting reservation has occurred; so UB passes the message to CWB.

(26) CWB creates a ConnectionAllowed message indicating the call can go ahead. It passes the message back to UB.

(27) UB sends the message to UD.

(28) (29) D's POTS Service-Agent (PD) takes the message and creates a message to the

Connection Agent (Cn) to set up bearer capabilities as in the basic call example. PD passes this new message back to UD.

(30) UD sends the message to Cn, which then proceeds to establish the bearer channel.

(31) The bearer channel is established.

(32) (33) Cn notifies UD, which notifies UB, that the bearer channel is established.

Note: (34), (36), and (38) may occur in parallel with (35), (37), and (39).

(34) (36) (38) UD has a POTS-supplied rule expressing interest in the notification from Cn, so UD passes the message to PD. PD creates a message to TD asking it to supply far-end alerting, and then passes that message back to UD. UD passes the message to TD.

(35) (37) (39) UB has a CW rule expressing interest in the notification from UD (this rule takes precedence over the corresponding POTS rule), so UB passes the message to CWB. CWB creates a message to Tx requesting CW-alerting and passes it to UB. UB passes the message to Tx.

5.3. Addressing Twenty-two Known Cases of FI

We have explained how the proposed architecture works in some relatively basic scenarios. We now consider how it could handle an example known case of FI.

The following example is taken, with minor changes in terminology, from the classic 'benchmark' paper [10]. In fact, our architecture handles all of the benchmark examples. We present here one illustrative example:

5.4. CallWaiting (CW) and AutomaticCallBackLater (ACB) - Benchmark Ex. 13

User A subscribes to CW. User B subscribes to ACB. If B calls A while A is already on the line, then ACB might be fooled into thinking the line wasn't busy, since CW causes B to receive ring-back instead of busy-tone.

(Numbers refer to Figure 4.)

Note that the above behavior might be correct, if A doesn't want callers to know when he is ignoring incoming calls while already on the line. However, we assume here that the correct behavior is for ACB to take action.

A's User Agent (UA) fields the ConnectRequest by passing it to A's POTS Service-Agent (PA), as in earlier examples. PA creates a message requesting terminal resources. This message gets passed via UA to the Terminal Agent (TA) for A's current terminal. TA informs UA that the terminal is busy. UA has a rule supplied by A's CW Agent (CWA), expressing interest in such busy-notifications, so UA passes the notification to CWA. CWA creates a request for TA to reserve functionality, including a CW alert, for B's call. UA passes the request to TA, which reserves alerting functionality. The interchange between Agents continues as in the basic call scenario, except at the end where CWA once again has supplied a rule expressing interest, this time in the message from the Connection Agent(s) (Cn) that the call is set up. CWA creates a message to TA to apply CW alerting. The message is passed on via UA, and the call reaches the state where A receives CW-alerting while B receives far-end alerting. Thus ACB (correctly) stays dormant in this case.

In contrast (Figure 5), consider the situation where A doesn't subscribe to CW. Then when UA receives the Busy message from TA, no one but PA has supplied rules expressing

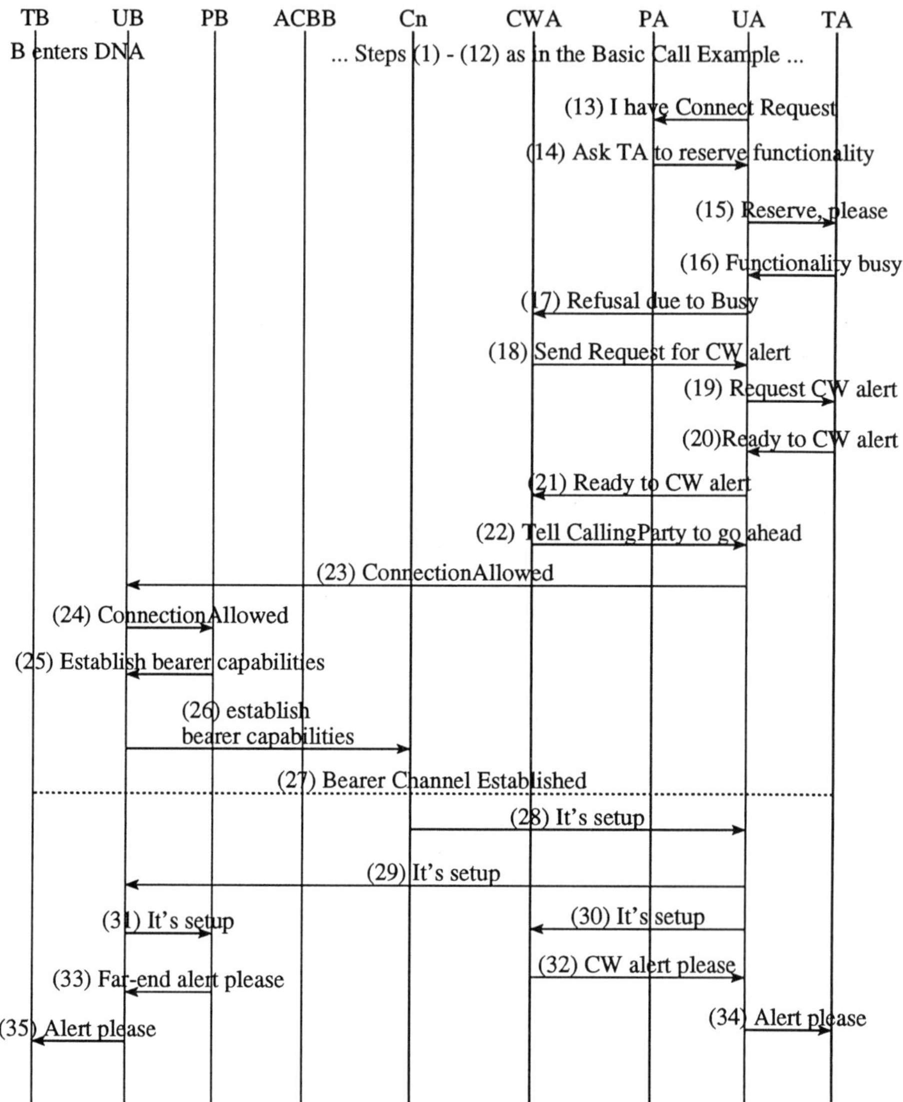

Figure 4: CallWaiting and AutomaticCallBackLater (Benchmark Ex. 13)

interest. PA constructs a Busy message to go to UB. ACBB has supplied a rule expressing interest in that message, and prepares to call back later (perhaps by establishing a rule asking for notification from TB about the current call being cleared).

5.5. OperatorServices and OriginatingCallScreening (OCS) - Benchmark Ex. 6

User A subscribes to OCS (installed at A's local switch). An operator in a remote switch doesn't know about A's services, and so treats all calls made by A as POTS. Thus A could

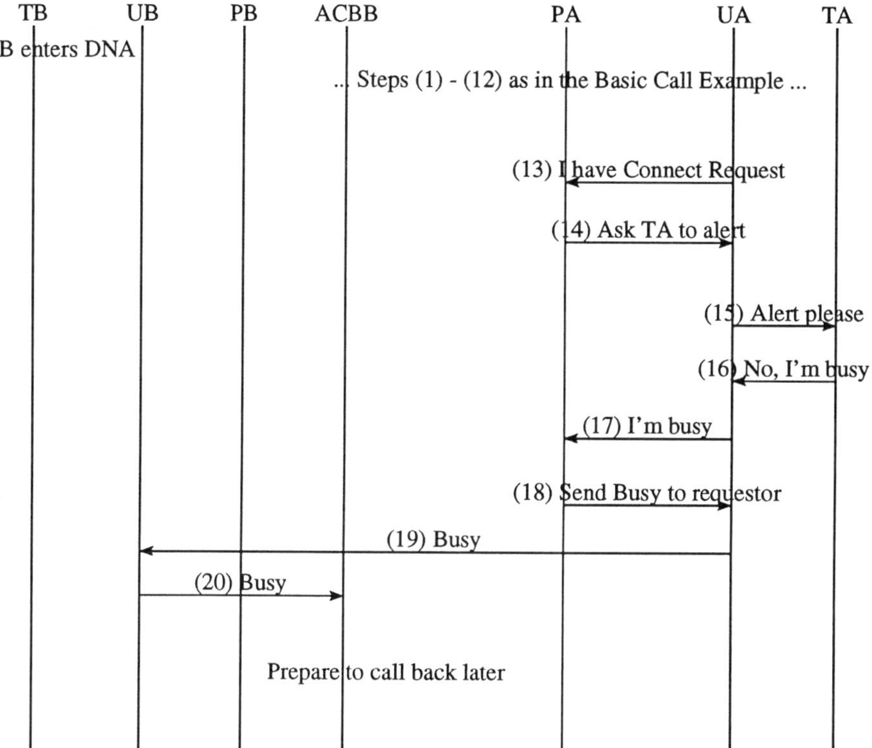

Figure 5: CallWaiting and AutomaticCallBackLater (Benchmark Ex. 13) Variation
with no Call Waiting

make an operator-assisted call to a DN disallowed by OCS.

(Numbers refer to Figure 6.)

When A asks the operator for B's DN the operator, acting just like A's terminal, gives a ConnectRequest to A's User Agent (UA) with the DN set to B's DN. A's POTS Service-Agent (PA) creates a ConnectToAgent request with the DestinationAgent parameter set to B's User Agent (UB), as in the basic call scenario. UA has a rule supplied by A's OCS Service-Agent (OCSA), expressing interest in the ConnectToAgent request; so OCSA examines the DestinationAgent parameter and reacts by generating a request-denial, which UA sends back to the operator. If the operator respects this denial, then the call does not succeed.

We could envision at least two ways to let the operator optionally override OCS: The ConnectRequest and ConnectToAgent request could have OverrideScreening parameters, which cause OCSA not to fire its rule (depicted in Figure 6). Or, the operator could simply establish the call to B via Connection Agents.

6. Areas for Further Investigation

This paper has addressed FI in the traditional narrowband telephony environment. Addi-

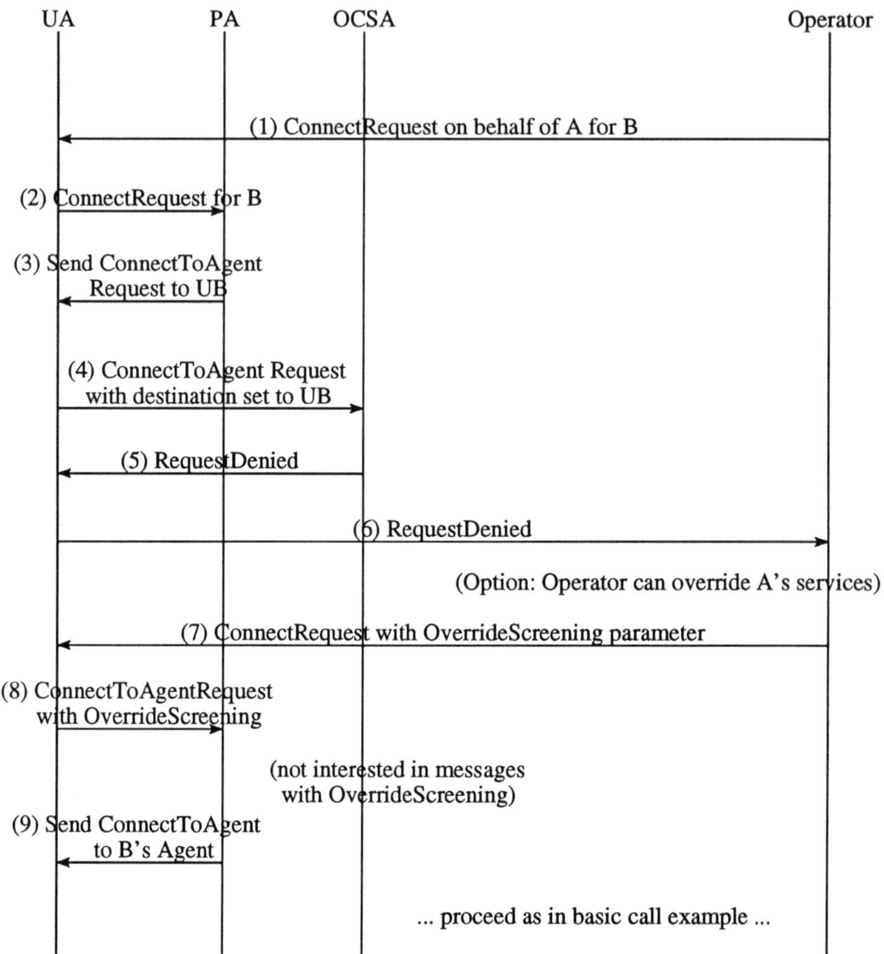

Figure 6: OperatorServices and OriginatingCallScreening (Benchmark Ex. 6)

tional work is needed to investigate the implications of our approach with respect to legacy systems, current IN models and emerging broadband models. Our approach will also benefit from further investigation and refinement in areas such as: mechanisms for logical/physical signal binding, charging, describing service intent with dispatch rules, dealing with detected ambiguities and the level of granularity required in parameter/event/operation definitions. While we feel our approach is already fundamentally sound, we expect it to evolve as a result of further work.

7. Conclusions

It is extremely difficult, if not impossible, to control feature interactions in an environment which does not allow ample manipulation of the basic objects of service processing and which does not allow expression of differences among such objects. In particular, the environment must be designed with the expectation that new services will be created, new resources will be made available, and existing actors will be asked to play new roles. The current environment for narrow-band services does not, by itself, ensure that new services can be added without substantial reworking of older services.

We have presented an architecture based on separation of concerns, supported by a processing model to provide message and event handling. We claim that combining these meets the needs for an extensible system of new services and technologies. To support these claims we have shown how the architecture and processing model allow solutions of reported feature interactions for narrow-band telephony services and universal personal telecommunications services.

Our architecture is heavily influenced by the goal that the introduction of new services and new technology should not require modifications of existing software. Our separation of concerns approach allows us to come close to that goal, while our processing model enables a service independent method of managing interactions. In many, we hope most, cases in which the system is unable to resolve a conflict we provide detection mechanisms (logging of precedence ambiguities and of possible role misunderstandings) which should help pinpoint exactly where the modifications are needed.

It is difficult to totally validate an architecture which claims to help avoid problems which no one has ever seen. It is still possible, perhaps even likely, that the proposed architecture embodies assumptions which will preclude easy integration of some unforeseen types of services. Nevertheless, our experience thus far indicates that this architecture has necessary and desirable characteristics and is a worthy candidate as an infrastructure for telecommunications services.

Bibliography

[1]   Bowen, T.F. et al., The Feature Interaction Problem in Telecommunications Systems, *Proc. of the 7th IEE Int. Conf. on Software Engineering for Telecommunications Systems,* July 1989, p. 59-62.

[2]   Cameron, E.J. and Veltjuijsen, H., Feature Interactions in telecommunications systems, *IEEE Communications Magazine,* August 1993, p. 18-23.

[3]   Zave, Pamela, Feature Interactions and Formal Specifications in Telecommunications, *IEEE Computer,* August 1993, p. 20-30.

[4]   Bouma, L.G. and Veltjuijsen, H., Introduction to *Feature Interactions in Telecommunications Systems,* IOS Press 1994, p. vii-xiv.

[5]   Barr, W.J. et al., The TINA Initiative, *IEEE Communications Magazine,* March 1993, p. 70-76.

[6]   van der Linden, R., Using an architecture to help beat feature interaction, *Feature Interactions in Telecommunications Systems,* IOS Press 1994, p. 24-35.

[7]   Lauer, Gregory S., IN Architectures for Implementing Universal Personal Telecommunications, *IEEE Network,* vol. 8, no. 2, March/April 1994, p. 6-16.

[8]   Collins, R., Harris, S., Willis, D., Feature Interaction and AIN Evolution, *Proceedings TINA 92,* Narita, Japan, January 1992.

[9]   Cross, M. and O'Brien, F., Restructuring the Problem of Feature Interaction: Has the Approach been Validated? Experience with an Advanced Telecommunications Application for Personal Mobility, *Feature Interactions in Telecommunications Systems,* IOS Press 1994, p. 249-257.

[10]   Cameron, E.J. et al., A Feature Interaction Benchmark for IN and Beyond, *Feature Interactions in Telecommunications Systems,* IOS Press 1994, p. 1-23.

Feature Interactions in Telecommunications Systems III
K.E. Cheng and T. Ohta (Eds.)
IOS Press, 1995

An Interaction-Avoiding
Call Processing Model

D. Cattrall, G. Howard, D. Jordan, S. Buj
BNR Europe Limited, London Road, Harlow, Essex, CM17 9NA
tel: +44 1279 429531   fax: +44 1279 441551 email: {D.M.Cattrall, G.Howard}@bnr.co.uk

Abstract. This paper describes a new call processing model that is designed to avoid feature interactions. This model uses novel concepts to address known causes of interactions and provide a platform that allows users to better express their intentions. The concepts used in this model are described and an experimental implementation derived by rapid prototyping techniques is discussed. We present the results of initial experiments that show how the model avoids certain common feature interactions.

1. Introduction

The feature interaction problem and approaches to dealing with it are well documented in the literature. For example, the proceedings of the second workshop on feature interaction [1] describes many approaches to detecting, resolving and avoiding interactions.

Cameron [4] presents a taxonomy of feature interactions and categorises these causes:

Assumptions about Naming: Historically a telephone number uniquely identified a telephone line. However, modern features have made this 'one number–one line–one entity' model obsolete. Different features make different assumptions about whether a telephone number names the physical line, the individual(s) using that line, or even the associated service.

Assumptions about Call Control: Call control refers to the ability of a user to manipulate a call. Interactions arise when the call control of one feature prevents other features from exercising their control. This problem is likely to arise if one party has control that is not expected by other parties.

Assumptions about Data Availability: Features cannot work properly unless they are able to obtain the necessary data. Some features assume certain kinds of data should be available, others assume it should be private.

Assumptions about Administrative Domain: An administrative domain is a telephone network administered by a single organisation. A feature that is defined in one administrative domain may not be usable from another administrative domain.

Assumptions about Signalling Protocol: Interactions occur when a feature needs to know the status of the called party but cannot correctly tell what the status is from the signal received.

Limitations on Network Support: Network components have their own limited capabilities in communicating with other network components or processing calls, which in turn can result in feature conflicts.

Intrinsic Problems in Distributed Systems: Telecommunications systems are huge, real-time, reactive distributed systems. Many difficulties in dealing with large distributed systems are also present in managing feature interactions.

1.1. Outline of our Approach

We propose a new call processing model that attempts to avoid feature interactions. It uses novel concepts to provide a platform that allows users to better describe their intentions. Moreover, these concepts are designed to avoid interactions by addressing the causes cited above.

Some of these causes are resolved directly. Assumptions about naming are avoided by drawing a distinction between the identity of a piece of equipment and the address of a user. Assumptions about call control are avoided by stating the control of each party at call set up.

We take a high-level view of telecommunications since we are interested in detection of interactions caused solely by the definitions of features. Therefore we resolve other causes by making simplifying assumptions about the underlying network:

- There is just one administrative domain.
- There are no signalling limitations.
- The data required for a feature's functionality is always available.
- There are no problems in the underlying network or in its support.

2. Components of the Call Processing Model

Figure 1 illustrates the relationships between the components that we believe are needed to describe call behaviour. Solid lines denote relationships that are manipulated during call processing; broken lines to the user indicate other conceptual relationships. The components are described below.

2.1. Access Components

In conventional networks it is not clear whether a telephone number refers to the call participant, his life role, or the equipment he uses. We introduce separate terminology for each of these concepts.

2.1.1. Roles

Each role denotes a user or group of users of the telecommunication network. Each role has a unique identifier. Role identifiers are used to identify parties within a call.

People assume different *life-roles* for which they have different communication needs. The use of roles in call processing is intended to reflect these needs so that calls can be handled accordingly.

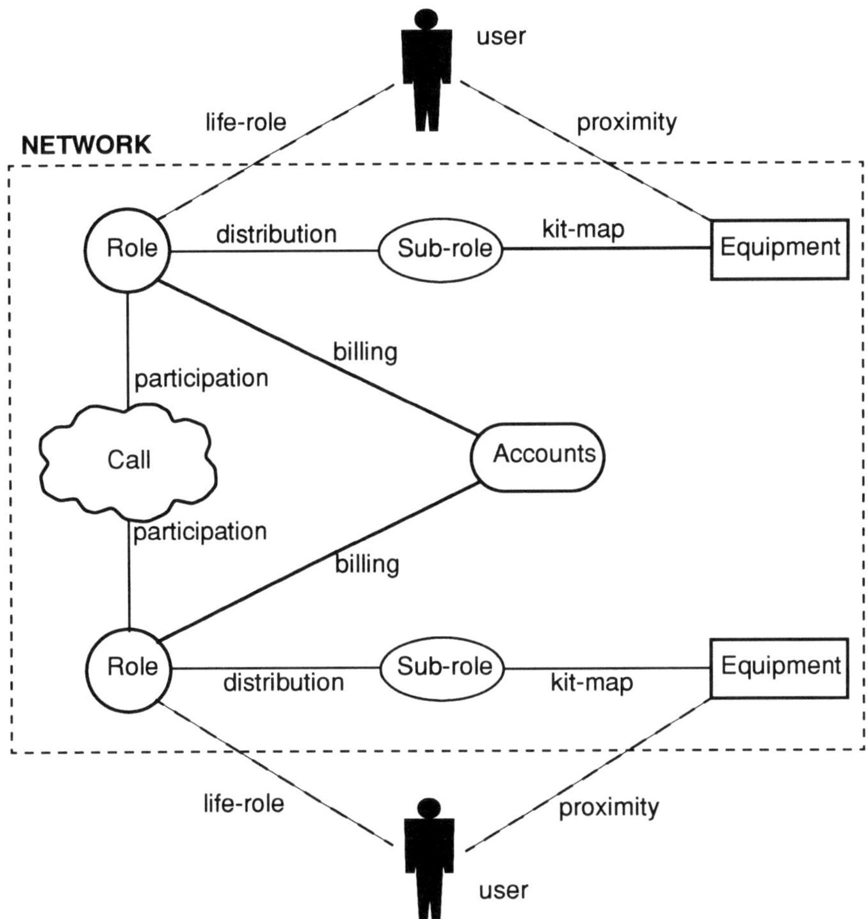

Figure 1　Components of the call processing model

For example, a person may have a work role for his job, a first-aider role for his job, a private/personal role, and a role as a golf club captain. Undesirable interactions are greatly reduced by making a distinction between each of these roles. Providing the equipment that person is using can indicate which of his roles is being called, he can answer appropriately or even choose to ignore certain roles at certain times of the day etc.

This system allows greater flexibility than is currently possible. For example, the golf club captain role can be re-assigned to a different person each year as a new captain is elected. There is no need to redistribute personal details on how to contact the current captain: contact will remain the same via the club captain role. Only the association between this role and the equipment to which its calls are directed needs to change.

2.1.2. Sub-roles

Sub-roles represent the different ways in which a role might handle its calls. In particular, sub-roles are used to describe call *distribution*. Calls can be distributed to different people or to different equipment types by representing them as distinct sub-roles within the role.

Different sub-roles of a role may behave differently depending on individual preferences and equipment capabilities.

Each sub-role may be associated with several different items of equipment. Each call must be proposed from a specific role and sub-role. The equipment used to initiate or receive a call must be associated with the involved sub-role.

2.1.3. Equipment

Each item of equipment has a location and indicates which channels are currently being used and which are available. Several roles may be associated with the same item of equipment through their sub-roles.

The identities of pieces of equipment are hidden from the call set up process. Instead there is a *kit map* that relates each role to the item(s) of equipment that role currently uses.

The association of equipment with a sub-role by the kit map is intended to be flexible and is potentially dynamically updatable. As a person moves around he may modify the kit map to identify the piece(s) of equipment that should be associated with his sub-roles.

For example, the updatable kit map can be used to model the dynamic association that happens automatically in a mobile network. Personal communication services and universal personal telecommunications can be modelled similarly.

Another use of roles and sub-roles is to represent the concept of a hunt group. A hunt group has a role that is called when the caller wishes to reach any member of a department without having to know the address of a particular member. Each member of the department is denoted by a sub-role. Any incoming call received by the department role is distributed to one of these sub-roles. "Call Filters and Preferences" on page 5 describes how the user can influence this distribution.

2.2. Call Related Components

2.2.1. Calls

Calls are the basic units of communication. A call consists of one or more *participations*. Each participation denotes the involvement of a particular role represented by one of its sub-roles, using a connection of a particular type (e.g. voice, video, audio or still image) with a direction (transmit or receive), a start time (set when the call is connected) and a finish time (set when the call is released). Constraints such as duration limit, cost limit etc. can be set on each participation. For example, total cost constraints may be defined to model callbox services.

To complete a call, each participation is assigned equipment that is associated with the sub-role by the kit map and which supports the required type of connection. It is the task of the network to ensure the proper inter-connection of participations.

For future systems it will be desirable to handle multi-party calls. We model these calls by having one participation for each party.

2.2.2. Users

Users are ordinary people. They initiate and receive calls in a particular role and sub-role. Each user may have sub-roles in one or more roles. He can use various items of equipment associated with his sub-roles by the kit map. The kit map denotes the implicit notion of *proximity* between user and equipment.

2.2.3. Accounts

Accounts contain all call related *billing* information. This includes role subscription charges, bandwidth usage and transaction (or feature) based charges. Accounts are uniquely associated to roles.

Charges for bandwidth usage are allocated to participations in a flexible manner so that call charges are treated as an intrinsic property of the participation. For example, freephone calls can be represented by defining no charge on caller participations and an appropriate charge rate on the callee participations.

3. Call Processing

Using the concepts introduced above, call processing methods that reduce the potential for feature interactions can be defined. This approach incorporates some traditional features as a fundamental part of call set up. Call distribution is intrinsic to the description of participations, and a wide range of call screening behaviours can be achieved through the use of call filters and preferences.

This provides a basis for the definition of call management services that have a reduced likelihood of undesired interactions.

3.1. Services

We regard a service as being the package of telecommunication operations available to a role. Each of these operations may be regarded as a feature. When a user subscribes to a service he is given a role, which behaves in accordance with the operations defined in that service. Users can only invoke operations belonging to the services to which they subscribe.

3.2. Call Filters and Preferences

The user can define incoming and outgoing call filter functions for each sub-role. An outgoing call will only be accepted from a sub-role if permitted by its outgoing filter (this models the originating call screening feature). When a role receives an incoming call, the incoming filter of each of its sub-roles determines whether that sub-role is prepared to accept the call (this models the terminating call screening feature). The call must either be distributed to one such sub-role or else rejected. The user can control this distribution by defining preferences amongst sub-roles.

Each filter can test all aspects of a call and its participations – the roles and sub-roles involved in the call, the time and duration of the call and even the cost. This allows sophisticated filters to be defined.

The network operator can also specify a filter that determines which calls the network is prepared to connect. For example, this may be used to control access to and from a virtual private network.

3.3. Proposals and Call Selection

Each call is initiated as a *proposal*. A proposal describes all of the alternative calls that could be set up between the caller and callees depending on the services to which they subscribe. Methods for proposal formation have not yet been fully investigated but we envisage the proposal being formed by agreement between services. The use of proposals allows alternative forms of connection to be made if the preferred one is not possible.

Each call in the proposal is tested against the acceptance criteria specified by the filters of the participating sub-roles and the network filter. Also the equipment associated with each sub-role is tested to determine whether it is available. For a call to be connected the participating sub-roles must be prepared to accept it and also have equipment available. Any call that fails to pass one or more of these tests is removed from the proposal.

If no call in the proposal is acceptable then no connection is made and a rejection explanation is given. If one or more calls in the proposal is acceptable then one of them is selected for connection. To assist this decision each participating sub-role may specify incoming and outgoing call preferences. However, this may not be sufficient to identify one unique call for connection, in which case it is necessary to make a final choice amongst the remaining alternatives. Network operator preferences may be taken into account when making this choice. The combination of incoming/outgoing participant filters, network filter, incoming/outgoing participant preferences and final network operator choice provides a method for *call selection.*

In each call of a proposal each participation has the same start time – the time at which the initiator would ideally like the call to start. This permits creation of an *active proposal*, in which the start time is `now` (indicating immediate connection is required), or a *deferred proposal*, in which the start time is either a future value or `earliest` (meaning connect at the earliest possible time). When the appropriate time for an active or a deferred proposal is reached, an attempt will be made to connect one of the calls. If a call is connected then the start time of each participation is set to the current time.

Deferred proposals act as the basis for modelling pre-booked calls, call-back features and queuing features. For example, the automatic call-back feature can be modelled by creating a deferred proposal in which the start time is `earliest`. If the callee's equipment is currently in use then the call will be connected when that equipment becomes available.

3.4. Call Control

One form of call control is call release, whereby a person attempts to remove his participation in a call. Our call processing model has several different methods of release:

release-all. This shuts down the whole call releasing all the participants. For example, in a two-party call both parties would typically have release-all. If either releases his participation then the call is terminated. Similarly, the initiator of a conference call would use release-all if he is paying the bill.

simple-release. This releases the participant from the call but allows the call to continue with all other participants communicating as before. A person who is participating in, but is not the initiator of, a conference call may use simple-release.

no-release. This means that the participant cannot release himself from the call. This may be used to model emergency service[1] calls.

release-choice. This is used for those circumstances where the release method cannot be determined at call proposal formation. For example, a person initiates a two-party call, which would typically imply release-all — if he releases then the call is shut down. However, if the intention is to include a third person in a conference call then release-all may be deemed inappropriate.

Many feature interactions ensue from one call participant modifying the call in a manner that is unexpected by other participants. As Cameron [4] observes, such behaviour is unexpected because the call control of each participant is not explicitly stated.

Conceptually, a participant can modify a call if he can terminate that call and initiate a new call with the required modification. Currently, we use this mechanism to model call modification. However, the only way for a participant to terminate a call is to release his participation. Thus the release method available to a participant determines the extent of his call control. By insisting that the release capability of each participant is stated when the call proposal is formed we avoid assumptions about call control.

4. Experimental Implementation

4.1. Implementation of the Call Processing Model

Our model would be referred to as an information model in Open Distributed Processing (ODP). Following previous ODP work [2][3] we use Z notation [7] for modelling.

---

1. The traditional example is the U.S. emergency 911 service

Having formally specified our model in Z, a rapid prototype was developed in the functional language Standard ML (SML) [8]. The Z specification provided a rigorous framework for discussion and development of our model, while the SML prototype allowed experimentation. As a result of discussions and experimentation successive versions of our model were specified and implemented.

SML was chosen as the prototyping language because it allows programs to be expressed in a notation that is conceptually very close to Z, which is based on typed naïve set theory, predicate calculus and schema calculus. SML provides lists and abstract types, which together can be used to model finite sets, and hence finite relational/functional mappings. Predicates and schemas are easily translated into SML functions and schema calculus operations translated into higher-order functions. The notion of state used in Z operations can be represented in SML as references (updatable storage).

SML provides modules that have clearly defined interfaces, which encourage good software engineering practice. The operations that change and access the state of our model form the call processing module. Services can be built as additional modules. Features can be incorporated into services as new operations or as modifications to existing operations. This use of modules permits a natural distinction between call processing, services and features.

We believe that a graphical interface greatly assists experimentation with a model of this nature but SML has no graphical capability. Therefore, although our SML prototype can be used as an experimental tool in its own right, we linked it to a graphical interface using a commercial graphical display system.

4.2. Implementation of Services and Features

We have modelled experimentally the services described below. Whenever a user subscribes to one of these services he is assigned a role and one or more sub-roles. The user specifies the incoming and outgoing filter and preference functions for these sub-roles, as required.

Basic voice call. This provides operations for set-up and release of basic calls involving voice connections. The subscriber is assigned one role and one sub-role with which to initiate and receive calls.

Basic video call. This provides operations for set-up and release of basic calls involving voice and video connections. The proposal formation operation caters for either a combined video-voice call or a voice-only call. This allows the subscriber to communicate with a person whose equipment supports voice, but not video connections. In this case a voice-only connection is set-up.

Manager-secretary. This provides the subscriber with a role that has two sub-roles — one for the manager and one for the secretary. Incoming calls can be received by either sub-role subject to their incoming filters. Outgoing calls can only be made by the manager sub-role and are set up in the same way as the basic voice call service.

Hunt group. This provides the subscriber with one role that has a specified number of sub-roles. The role represents a group and the sub-roles represent members of that group. Incoming calls are routed to the first available sub-role in a round-robin fashion. It is not possible to make an outgoing call from any of the sub-roles.

Universal personal telecommunication. This provides a 'here-I-am' operation that allows the subscriber to change the items of equipment associated with his role. The operations for set-up and release of basic calls are the same as the basic voice call service.

5. Experimental Results

Experiments with our model have shown that our prototype services can interact in the intended manner. "Examples of Feature Interactions Avoided" on page 8 demonstrates how our model avoids certain common feature interactions identified by Cameron[4]. "Example of a Feature Interaction Not Avoided" on page 9 gives an example of an unwanted feature interaction that persists. We consider possible solutions to this interaction.

5.1. Examples of Feature Interactions Avoided

It is possible for the conventional call forward feature to generate loops. For example, a person re-directs all incoming calls to his own number; or calls from A are forwarded to B, who forwards to C, who forwards back to A. This cannot happen in our model because there is no conventional call forward, instead there is call distribution to one of a finite number of sub-roles.

Further problems with call forward arise because the user's intention is not clear. As Griffeth and Velthuijsen [5] observe there are two main reasons why a person may use this feature:

1. To make sure that calls still reach the person when he is elsewhere. In our model this can be accomplished by associating new equipment with sub-roles in the manner of the universal personal telecommunication (UPT) service.

2. To redirect calls to someone else who can handle the call, e.g. a secretary. In our model this can be accomplished through use of roles and sub-roles. Each person that may possibly be required to answer the call can be assigned a sub-role. The hunt group and manager-secretary services were designed to handle this behaviour.

Another problem can arise with conventional features when calls are forwarded to a number in the outgoing call screening list. For example, parents block calls to a premium rate chat-line number from their line. Their child can circumvent this by forwarding incoming calls to the screened number and then calling himself!

In our model a unique role identifier must be specified for communication with the chat-line service to take place. The parents' outgoing filter will prevent calls from being connected to this role. As there is no conventional call forward in our model there is no way in which the child can communicate with this role.

Many of the feature interactions associated with call screening occur because it is unclear what is being screened, whether it is the actual telephone number, the person or service associated with that number. Our model avoids this by making it clear that roles, sub-roles, charge, time of day etc. are being screened, not the equipment number.

5.2. Example of a Feature Interaction Not Avoided

Our experiments have shown that while unexpected interactions between services in a call seem to have disappeared, problems remain in integrating distinct services for users with multiple roles.

For example, a project group subscribes to the hunt group service. Each member of the group is denoted by a sub-role and has access to a unique phone. One of the group members, S, also subscribes to our universal personal telecommunication service — this is his work role identified as 'S-work'. Therefore, when S moves to a different office he may use the UPT service to update the kit map so that his S-work role is now associated with the phone in that office. However, when he moves, the equipment associated with his project hunt group sub-role will remain unchanged. The hunt group does not inherit the mobility of universal personal telecommunication.

This problem cannot be easily solved because we have lost sight of the intention behind the combination of these two services. There is more than one possible resolution:

1. The project group could subscribe to a universal personal telecommunications hunt group service. However this would give all sub-roles mobility, which may not be the intended behaviour.

2. A specialised version of the hunt group service that gives mobility to just the sub-role associated with S could be created. By using separate 'here-I-am' operations for the S-work role and group sub-role, S could update the kit map to track his movements. However S may find multiple 'here-I-am' operations confusing and inconvenient.

3. The hunt group service could hunt across the services provided for each member. The hunt group sub-role for S would be equivalent to the S-work UPT role, while the sub-roles of the other members would be their basic voice call roles. However, service creation methods would have to be improved for services to be combined in this manner.

The way ahead can only be decided in the context of a general approach to service creation, which we intend to consider as part of our future work.

6. Related Work

Van der Linden [9] describes an architecture that is designed 'to help beat feature interaction'. His paper is based on results of work on open distributed processing. It advocates the principles of *separation* and *substitutability* for application independent structuring of systems. We also use the principle of separation but for information modelling rather than computational modelling.

Griffeth and Velthuijsen [5] describes an architectural approach using negotiating agents to resolve feature interactions. The negotiation is intended for use in conventional telecommunications systems where the user's true intent is not always clear. Consequently this mechanism must sometimes infer the user's intent. Ultimately we intend to incorporate a similar mechanism in our model to assist proposal formation and avoid the need for inference by offering a platform that better expresses user intent.

Collins et al. [6] propose an architectural model based on agents, which helps separate the network user from equipment in a similar manner to roles in our model.

Cross and O'Brien [10] argue that the problem of feature interaction can be recast as a realisation of the more abstract patterning of communication from the perspective of social interaction. To validate their approach they designed and prototyped an advanced telecommunications application that utilises additional parameters of time and media to avoid feature interaction. They conclude from their work that feature interaction can be avoided at the highest level by re-representation. This stance is close to ours — the aim of roles is to better express user intent at a high level.

7. Conclusions

We have presented a new call processing model that is based on several novel concepts intended to resolve the causes of feature interactions. Moreover, we believe these concepts provide a framework that better expresses user intentions

Our model provides a means of distinguishing the network users from the equipment they use via roles and sub-roles.

Interactions caused by one party having unexpected call control are avoided by stating the control of each participant at call set up.

The incoming and outgoing call filters associated with sub-roles provide a powerful and flexible call screening mechanism that screens roles, sub-roles, etc., i.e. the network users, rather than the equipment they use. Call preference functions combined with call filter functions provide a mechanism for call distribution that permits definition of multi-user services such as the manager-secretary and hunt group.

8. Future Work

Our immediate aim is to demonstrate model stability — to show that it is possible to build more complex services on top of our model, add features to those services and avoid unwanted interaction between those services and features. We intend to investigate more sophisticated service creation methods.

Call control needs to be investigated further. We need to investigate who has the right to control calls and in what manner. Also more flexible call modification operations are required to model certain features and hence feature interactions.

Our model defines an abstract view of telecommunications that is intended to reflect end-user to end-user communication and therefore has some simplifying assumptions that will have to be addressed:

- There is only one network. We realise that in practice an end-to-end call is routed across many networks (both public and private) but initially these may be viewed as a single network without either complicated billing and/or restrictions on what is possible.

- There is no delay in the call set up process.

- There are no physical limits other than the channel capacity of the end-user's terminal device.

- There are no notions of signalling or switching.

Once our model has been proven it will be necessary to demonstrate that it is feasible to implement in a real telecommunications network. By refining our model towards an implementation, compromising choices may have to be made. This could be a fruitful area for predicting feature interactions.

9. Acknowledgements

Special thanks to Will Harwood and Chris Paine of Imperial Software Technology for their contribution to the development of the call processing model and Z specifications.

Thanks to colleagues at BNR for many helpful comments on earlier versions of this paper.

10. References

[1] Feature Interactions in Telecommunications Systems" IOS Press, ed. L.G. Bouma, H. Velthuijsen, 1994

[2] Modelling Information Objects in Z" Open Distributed Processing IFIP, S. Rudkin, Elsevier Science Publisher 1992

[3] Object Oriented Modelling in Z for Open Distributed Systems" Open Distributed Processing IFIP, E. Cusack, 1992

[4] A feature interaction benchmark for IN and beyond" E. Cameron in [1]

[5] The negotiating agents approach to runtime feature interaction resolution" N.D. Griffeth, H. Velthuijsen in [1]

[6] Feature Interaction and AIN Evolution" Proceedings TINA Workshop, R.Collins, S.Harris, D.Willis, January 1992

[7] The Z Notation A Reference Manual" Prentice Hall International Series in Computer Science, J.M. Spivey, 1987

[8] ML for the Working Programmer" Cambridge University Press, L.C. Paulson

[9] Using an architecture to help beat feature interaction" R. van der Linden in [1]

[10] Restructuring the Problem of Feature Interaction" M. Cross, F. O'Brien in [1]

A Dynamic Resolution Method for Feature Interactions and Its Evaluation

Yoshiaki KAKUDA[†],Akihiro INOUE[†],Hiroyuki ASADA[†*],Tohru KIKUNO[†]
and Tadashi OHTA[††]

[†]Department of Information and Computer Sciences
Faculty of Engineering Science, Osaka University
1-3, Machikaneyama-cho, Toyonaka-shi, Osaka 560, Japan
Phone: +81 6 850 6566
Fax: +81 6 850 6569
E-mail: kakuda@ics.es.osaka-u.ac.jp
[††]Communications Software Department
ATR Communication Systems Research Laboratories
2-2 Hikaridai, Seika-cho
Soraku-gun, Kyoto 619-02, Japan
Phone: +81 7749 5 1230
Fax: +81 7749 5 1208
E-mail: ohta@atr-sw.atr.co.jp

Abstract. The Advanced Intelligent Network requires addition and modification
of communication services without invalid feature interactions between communication services. This paper proposes a dynamic method for resolving the feature
interactions. Since this method is based on the protocol synthesis technique, the
feature interactions are resolved in the protocol specification rather than in the
service specification. As a result, feature interactions between communication
services can be dynamically avoided in the sense that during execution of the
communication services a user associated with the communication services collects information on the other users and decides which communication service is
selected based on the information. In this paper, the proposed method is outlined
using an example of a feature interaction between the call waiting service and the
call forwarding service, and the proposed method is evaluated by applying it to
examples of feature interactions between the telephone services described in Bellcore and detected by ATR. The evaluation results show that the proposed method
increases flexibility on selection criteria of communication services.

1. Introduction

With diversification of requirements for communication services, a lot of research and
development of the Advanced Intelligent Network (shortly, AIN) have been done for at-
taining its quick adaptation to addition and modification of communication services.
When multiple communication services are triggered by a call, they are executed in
parallel, and a system state after their execution may generally include a contradiction. Then it is said that invalid feature interactions occur. On the contrary, when the

---

* He is currently with Enterprize Network Development Dept., Fuji Xerox Co., Ltd.

system state does not include any contradiction, it is said that valid feature interactions occur. Since this paper discusses resolution of invalid feature interactions, invalid feature interactions are simply called feature interactions hereinafter. There are high possibilities to introduce feature interactions by adding new communication services or modifying communication services. Resolution of feature interactions between multiple communication services are thus one of the most significant problems for the AIN [4][12].

This paper proposes the following method: whenever communication services are added or modified, they are formally described in the service specifications and automatically translated into protocol specifications using the protocol synthesis technique. If feature interactions are included in the protocol specifications, then only one of communication services is selected during their execution. Since this method enables a flexible addition and modification of communication services and resolves feature interactions not in the service specifications but in the protocol specifications, we call it dynamic resolution method.

Protocol synthesis to be used in the proposed method is to derive a protocol specification from a primitive sequence graph, and is one of the most crucial techniques in protocol engineering [13] for designing protocol specifications. A lot of protocol synthesis techniques based on the FSM model have been developed [3][16][17]. When two or more users send primitives simultaneously to processes through different SAPs (Service Access Points), nondeterminism occurs in the protocol specification such that (1) a process can send a message or receive another message at a time and (2) a process can receive a message or receive another message at a time. The proposed method is based on the automated protocol synthesis methods [7][9][11] for deriving protocol specifications which are free from protocol errors caused by the nondeterminism described above. Feature interactions are regarded as nondeterminism such that a process can send a message or send another message at a time, which is explained in the proposed method.

The proposed method consists of the following procedures: (1) service specifications are described by the STR(State Transition Rules) method, (2) sequences of service primitives are generated from the service specifications, (3) protocol specifications are derived from the sequences of service primitives using the protocol synthesis techniques [7][9][11], and (4) conditional branch functions for resolving feature interactions are inserted into the protocol specifications. The key idea and the details of the proposed method are presented in [8] and [10], respectively. This paper describes the outline of the proposed method and evaluates the proposed method by applying it to several practical telephone services described in [1] and detected in [19].

The rest of the paper is organized as follows. In Section 2, fundamental definitions on feature interactions are given. Section 3 defines the feature interaction problem to be discussed in this paper. In Section 4 and Section 5, a dynamic method for resolution of the feature interaction problem is proposed; Section 4 gives translation from the service specification to service primitive sequences, while Section 5 presents the resolution method based on the protocol synthesis technique. In Section 6, the proposed method is evaluated by telephone service examples. Finally, Section 7 concludes this paper with future research.

2. Definition on Feature Interactions

2.1 Communication Architecture

Figure 1 shows a model of communication architecture in which the feature interaction problem is discussed in this paper. The upper layer of the communication architecture consists of users while its lower layer consists of processes. We assume that there is one-to-one correspondence between users in the upper layer and processes in the lower layer. User $i(i = 1, 2, \cdots)$ exchanges primitives with process i through an interface between the upper layer and the lower layer called Service Access Point(SAP). Each pair of two processes sends and receives messages through a communication channel in the lower layer.

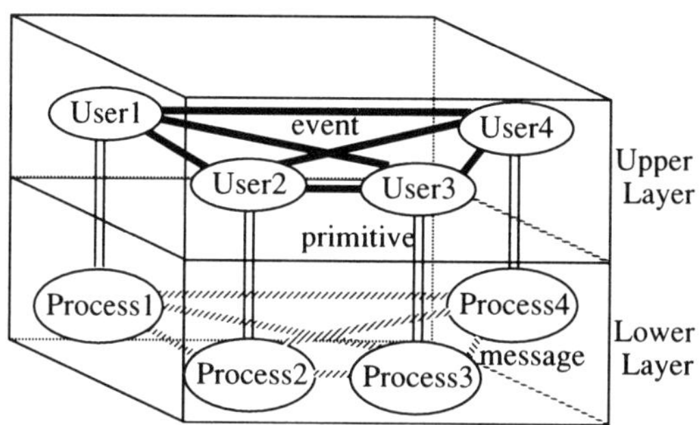

Figure 1: Communication Architecture

2.2 Service Specification

Since the upper layer consists of multiple users, a state of the upper layer can be composed by a set of states of such users. In the STR description method [5][6][18], a state of each user is represented by a collection of predicates. A service specification is described by service rules, each of which consists of (1) current states of users, (2) an event triggered by a user, and (3) next states of users. Two examples of service rules are shown in Figure 2.

The semantics of eight predicates on users in Figure 2 are explained as follows. Every predicate means the state of a user specified by the first argument.
[Example 1]

A predicate path(A,B) says that a connection for a call between users A and B is established. A predicate out-dialtone(C) says that user C can dial for connection of a

```
path(A,B), out-dialtone(C), m-cw(A)                    — current states
dial(C,A):                                             — event
out-ringback(C,A), cw-ringing(A,C), path(A,B), m-cw(A) — next states
```

(a) Call Waiting (CW) Service Rule

```
out-dialtone(C), path(A,B), idle(D), m-cfv(A,D)            — current states
dial(C,A):                                                 — event
out-ringing(D,C), out-ringback(C,D), path(A,B), m-cfv(A,D) — next states
```

(b) Call Forwarding Variable (CFV) Service Rule

Figure 2: Examples of Service Rules

call. A predicate m-cw(A) says that user A can receive a call from another user while user A is talking on the phone. A predicate out-ringback(C,A) says that user C is notified of the reception of acknowledge from user A by a ringback-tone. A predicate cw-ringing(A,C) says that user A is alerted to the reception of an additional call from user C by a side-tone. A predicate idle(D) says that user D is not talking on the phone. A predicate m-cfv(A,D) says that user A wants to activate a call-forwarding to user D. A predicate out-ringing(D,C) says that user D is alerted to the reception of a call from user C by a ringing-tone. Finally, an event dial(C,A) indicates that dialing from user C to user A is executed.

2.3 Feature Interactions

Feature interactions can be classified into several ones [14]. This paper discusses feature interactions by nondeterminism. The state of a communication system (simply, system state) is defined as a set of states of all users in the communication system. We say that states of users contradict with each other if there exist predicates representing the states of users which are satisfied by the system state and are incompatible due to semantics of the predicates. For a given system state and a service rule, if predicates representing the current states of users are satisfied by the system state, then the event in the service rule is said to be executable in the system state. If there exist two service rules including the same executable event in a given system state, then it is said that a feature interaction occurs for the two service rules in the system state and that execution of two events in the two service rules is nondeterministic. For given incompatible predicates, if the next states of users in the two service rules for which a feature interaction occurs contradict with each other, it is said that an invalid feature interaction occurs. Otherwise, it is said that a valid feature interaction occurs. In this case, such two nondeterministic events only have to be executed simultaneously. Resolution of an invalid feature interaction is to select one out of two nondeterministic events in the two service rules for which the invalid feature interaction occurs according to some criteria and to execute the selected event. Since this paper discusses resolution of invalid feature interactions, invalid feature interactions are simply called feature interactions hereinafter.

[Example 2]

Suppose that six predicates m-cw(A), m-cfv(A,D), path(A,B), path (B,A), out-dialtone(C), idle(D) are satisfied by a certain system state. Then, a feature interaction occurs for the CW and CFV service rules in Figure 2 because these two rules include the same event dial(C,A) that is executable in the system state. Since user C cannot try to connect more than one path simultaneously, out-ringback(C,A) and out-ringback(C,D) are incompatible with each other. The next states of the CW service rule and the CFV service rule thus contradict with each other and one of two dial(C,A)'s in the CW service and the CFV service should be selected.

2.4 Primitive Sequence Graph and Protocol Specification

Service primitives, shortly primitives, are transmitted and received between users in an upper layer and processes in a lower layer through an interface called SAP. The confirmation, indication, request and response (which are abbreviated as conf, ind, req and resp, respectively) are examples of primitives. A typical example of primitive sequence is req, ind, resp and conf. A collection of primitive sequences is represented by a labeled directed graph, which is called primitive sequence graph. In the graph, nodes, edges and labels represent states, transitions and primitives, respectively.

A protocol specification for process i is modeled by a Finite State Machine and represented by a labeled directed graph where nodes and edges respectively represent states and transitions and labels represent primitives, messages or ϵ's. A protocol specification is defined as a collection of protocol specifications for all processes. For easy understanding of a sequence behavior, the protocol specification is often represented by sequence charts, which are logically equivalent to the Finite State Machines.

Global behaviors of events or primitives in the system can be grasped from the service specification and the primitive sequence graph, while local behaviors of primitives and messages in each process can be understood from the protocol specification for each process.

3. Resolution of Feature Interactions

In this paper, the feature interaction problem is defined as follows:
Input $\cdots$ A service specification described by the STR description method.
Output $\cdots$ A protocol specification with capabilities in resolving the feature interaction.

The way to resolve the feature interaction problem is as follows. First, for each communication service in a given service specification, a sequence of primitives is generated from a service rule (Step1). Next, a protocol specification that may include feature interactions is derived from all the generated sequences of primitives (Step2). This derivation is done by the protocol synthesis methods in [7][9][11]. Finally, the feature interaction is resolved by inserting conditional branch functions to the protocol specification (Step3).

4. Generation of Primitive Sequences (Step 1)

An event in a service rule induces change in states of users, and this change is conveyed by primitives. When a feature interaction occurs, multiple events in service rules can become executable. In the proposed method, selection mechanism for multiple events is introduced in the protocol specification. Suppose that process S corresponds to a user which triggers the multiple events. The proposed method consists of the following two procedures.

(Procedure 1) Process S collects information on current states specified in the service rules by executing the first half of the primitive sequences, and based on the collected information, process S selects only one event out of multiple events.

(Procedure 2) Process S executes the latter half of the selected primitive sequence, and finally notifies users of their next states.

A distributed call processing architecture and modular signaling protocols are proposed in [2][15] for AIN. Unlike the conventional call processing, call control and connection control are separated and distributed, and feature interactions are resolved in the call control. The above (Procedure 1) and (Procedure 2) correspond to the call control and the connection control, respectively.

Criteria on the event selection in the proposed method are as follows.

(Criterion 1) Priorities are a priori given to all services and an event of multiple events in the rule of service with the highest priority is selected.

(Criterion 2) It is checked whether, for each service rule, all predicates in the current states of users are satisfied when their information is collected, and an event out of multiple events which are executable at that time is selected.

[Example 3]

Consider a case that higher priority is given to CFV service than that of CW service. In this case, process C collects information on current states in the CW and CFV service rules. It is easy to collect information on path(A,B), out-dialtone(C), m-cw(A), and m-cfv(A,D) by a typical sequence of primitives req, ind, resp, and conf concerning dial(C,A) and to check whether these predicates are true or not because dial(C,A) is an event from user C to user A. However, since idle(D) is a predicate in user D, it must be collected by another sequence of primitives through user D. Based on this collected information, the following decision is made: If dial(D) is true when information is collected, then the CFV service is selected according to (Criterion 1). Otherwise, the CW service is selected according to (Criterion 2).

Now, we can show an algorithm to generate primitive sequences in Figure 3. Without loss of generality, we suppose that $event(A, B)$ occurs and process A is taken as process S which is mentioned above. The sequence of primitives is generated as **print** sequence. A similar sequence of primitives is generated for an event with only one user and omitted.

```
variable Table K  /* a set of users associated with current states
                      specified in the service rules are stored in this table */

print("event_ req at SAP_A ↓");
print("event_ ind at SAP_B ↑");
print("event_ resp at SAP_B ↓");
Regist(A, K); /*   User A is stored to Table K                              */
Regist(B, K); /*   User B is stored to Table K                              */
for each U in current state /* which is the state of user U */ do
    if U ∉ K then /* If User U has never given its information to process A */
        print("state_ ind at SAP_U ↑")
        print("state_ resp at SAP_U ↓")
        Regist(U, K) /* Register user U to Table K */
for each U in current state do
    MakePreStateTable;
for each U in next state do
    MakePostStateTable;
/* Table (PreT(U) or PostT(U)) is made for each user U and               */
/* it has the set of the current(or the next) states of user U           */
for each U in post state do
    if PostT(U) ≠ PreT(U) and U ≠ A then
        print("change_ ind at SAP_U ↑")
        print("change_ resp at SAP_U ↓")
print("event_ conf at SAP_A ↑");
```

Figure 3: Algorithm to Generate Primitive Sequences

In this algorithm, SAP_A represents a SAP between user A and process A. Similarly, SAP_C and SAP_D represent such SAP's. For example, "event_ req at SAP_A ↓" denotes primitive event_ req which is delivered from user A to process A through SAP_A. "event_ ind at SAP_B ↑" denotes primitive event_ ind which is delivered from process B to user B through SAP_B.

dial_ req at SAP_C ↓ /* Process C knows $S_C{}^1$ */
dial_ ind at SAP_A ↑
dial_ resp at SAP_A ↓ /* Process C knows $S_C{}^1, S_A{}^1$ */ } collection of user states
cw_ ind at SAP_A ↑
cw_ resp at SAP_A ↓
dial_ conf at SAP_C ↑ } notification of user states

(a) Generation from CW Service Rule

dial_ req at SAP_C ↓ /* Process C knows $S_C{}^1$ */
dial_ ind at SAP_A ↑
dial_ resp at SAP_A ↓ /* Process C knows $S_C{}^1, S_A{}^1$ */
state_ ind at SAP_D ↑
state_ resp at SAP_D ↓ /* Process C knows $S_C{}^1, S_A{}^1, S_D{}^1$ */ } collection of user states
cfv_ ind at SAP_D ↑
cfv_ resp at SAP_D ↑
dial_ conf at SAP_C ↓ } notification of user states

(b) Generation from CFV Service Rule

Figure 4: Primitive Sequences

[Example 4]

By applying this algorithm to the CW service rule and the CFV service rule in Figure 2, we can get primitive sequences as shown in Figure 4. In this figure, $S_A{}^1, S_C{}^1$ and $S_D{}^1$ represent states of users A, C and D before execution of dial(C,A), respectively. Since the first half of primitive sequence in Figure 4(b) includes the first half of primitive sequence in Figure 4(a), event dial(C,A) can be executable in both cases. Thus, by inserting conditional branch functions between first half and latter half in each case, the feature interaction can be resolved.

5. Synthesis of Protocol Specification (Step 2 & Step 3)

Based on primitive sequences in which feature interactions occur, protocol specifications including transitions by transmission and reception of messages are derived using the following protocol synthesis methods in [7][9][11]. The procedure in the protocol synthesis methods is as follows.

(1) Based on a set of primitive sequences obtained in Section 4, a primitive sequence graph is formed.

(2) Dummy primitives are added into the primitive sequence graph. (Dummy primitives are primitives for correctly applying the synthesis rules to the primitive sequence graph.)

(3) The primitive sequence graph is projected into directed graphs representing each process.

(4) The synthesis rules in [7][9][11] are applied for each directed graph and transitions representing transmission and reception of messages between processes are generated.

(5) Dummy primitives and ϵ transitions are removed and a protocol specification is derived.

5.1 Primitive Sequence Graph

Two primitive sequences in which a feature interaction occurs for service rules i and j are generally represented as follows.

$$seq_i = (s_1{}^i, t_1{}^i, \cdots, s_p{}^i, t_p{}^i, \cdots, s_q{}^i, t_q{}^i, \cdots, t_{n_i-1}{}^i, s_{n_i}{}^i)$$
$$seq_j = (s_1{}^j, t_1{}^j, \cdots, s_p{}^j, t_p{}^j, \cdots, s_q{}^j, t_q{}^j, \cdots, t_{n_j-1}{}^j, s_{n_j}{}^j)$$

Here, $t_k{}^i (1 \le k \le n_i)$ and $t_l{}^j (1 \le l \le n_j)$ denote primitives for service rules i and j, respectively. $s_k{}^i (1 \le k \le n_i)$ and $s_l{}^j (a \le l \le n_j)$ denote states either before $t_k{}^i$ and $t_l{}^j$ are executed or after $t_{k-1}{}^i$ and $t_{l-1}{}^j$ are executed, respectively. Let $s_p{}^i (1 \le p \le n_i)$ and $s_q{}^j (1 \le q \le n_j, p \le q)$ denote states where information on a state of the communication system before the event in service rules i and j is executed is collected. If a feature interaction occurs for the service rules i and j, then the same events are executable. The first halves of the sequences seq_i and seq_j are thus equivalent and $s_a{}^i = s_a{}^j (1 \le a \le p)$ and $t_b{}^i = t_b{}^j (1 \le b \le p - 1)$ for all a and b.

A primitive sequence graph consists of the following primitive sequences, which is obtained from seq_i and seq_j.

$$seq_1 = (s_1{}^i, t_1{}^i, \cdots, s_p{}^i, t_p{}^i, \cdots, s_q{}^i, t_q{}^i, \cdots, t_{n_i-1}{}^i, s_{n_i}{}^i)$$
$$seq_2 = (s_1{}^i, t_1{}^i, \cdots, s_p{}^i, t_p{}^j, \cdots, s_q{}^j, t_p{}^i, \cdots, s_q{}^i, t_q{}^i, \cdots, t_{n_i-1}{}^i, s_{n_i}{}^i)$$
$$seq_3 = (s_1{}^i, t_1{}^i, \cdots, s_p{}^i, t_p{}^j, \cdots, s_q{}^j, t_q{}^j, \cdots, t_{n_j-1}{}^j, s_{n_j}{}^j)$$

Here, seq_1 represents a primitive sequence for service rule i when no feature interaction occurs. seq_2 represents a primitive sequence for service rule i when a feature interaction occurs. Note that seq_2 includes $t_p{}^j, \cdots, s_q{}^j$, which is a subsequence of seq_3. seq_3 represents a primitive sequence for service rule j regardless that a feature interaction occurs or not.

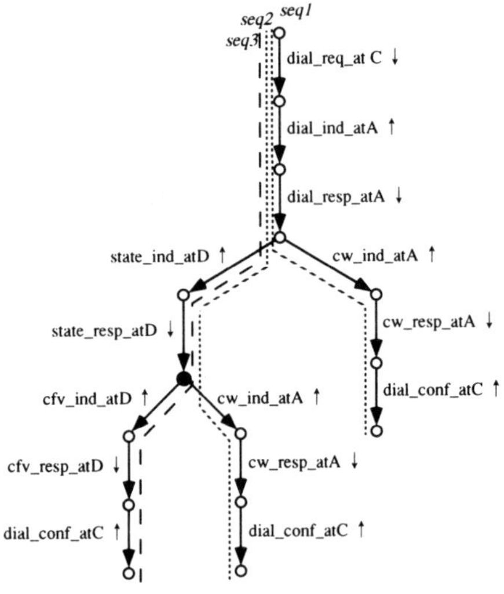

Figure 5: Primitive Sequence Graph

[Example 5]

Primitive sequences for CW service rule and CFV service rule are obtained as shown in Figure 4. Based on these sequences, a primitive sequence graph in Figure 5 is formed. In this figure, two sequences denoted by dotted lines represent seq_1 and seq_2 for CW service rule and a sequence denoted by a dashed line represents seq_3 for CFV service rule. At a state denoted by ● one of two events for which a feature interaction occurs is selected. In Figure 5, $dial\_reqatSAP_C \downarrow$ is denoted by $dial\_req\_atC \downarrow$ for concise description. The other primitives are similarly denoted.

5.2 Insertion of Dummy Primitives

Deriving a protocol specification from a primitive sequence graph is known as the protocol synthesis problem. A lot of methods for resolving this problem have been proposed so far. This paper adopts protocol synthesis methods proposed in [7][9][11]. These methods can derive a protocol specification from a primitive sequence graph, in which nondeterminism occurs when two or more users send primitives simultaneously to processes through different SAPs. In particular, the method in [11] enables an automated synthesis of multi-party protocols which consist of $n(n \geq 2)$ processes.

Before applying the above protocol synthesis methods, dummy primitives (denoted by dum) are inserted in the primitive sequence as follows.

(1) Before all edges leaving the branching node, insert a dummy primitive dum from user to process through SAP_j. As a result, one of primitives leaving the branching node can be executed.

(2) Insert a primitive dum from user j to process j between the primitive from user i to process i and the primitive from user k to process k. As a result, a message can be delivered from process i to process k through process j.

[Example 6]

Figure 6 shows the primitive sequence graph in which dummy primitives are inserted into the primitive sequence graph of Figure 5. In Figure 6, gray arrows denote dummy primitives.

5.3 Projection to Each Process and Generation of Messages

The primitive sequence graph with inserted dummy primitives expresses behavior of primitives concerning all $SAPs$. On the other hand, protocol specification expresses behavior of both primitives and messages concerning each process. To derive such protocol specification, the primitive sequence graph is projected onto each process. That is, for each process i, a directed graph is obtained from the primitive sequence graph so that primitives through SAP_i is left as they are and the other primitives are replaced by ϵ. We say this graph the primitive sequence graph for SAP_i. This graph expresses the dynamic behavior of primitives concerning only SAP_i.

In order to derive a protocol specification from the primitive sequence graph for each process, based on each successive primitives in the graph, messages to be delivered between processes are generated using the synthesis rules in [7][9][11].

5.4 Elimination of Transitions of Dummy Primitives and ϵ's

Finally, transitions of dummy primitives and ϵ's are eliminated from the protocol specification for each process. This elimination is done in the primitive sequence graph by combining both ends of edges labeled dummy primitive or ϵ and deleting the edge itself. The protocol specification in Figure 7 is derived from the primitive sequence graph in Figure 6 by projection of the primitive sequence graph to process A, generation of

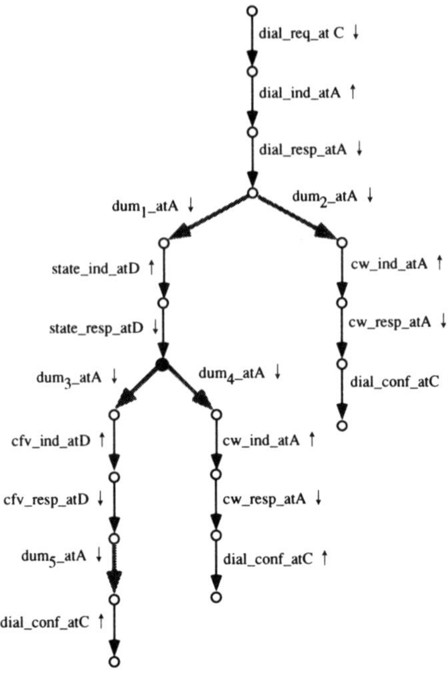

Figure 6: Insertion of Dummy Primitives

messages, and elimination of dummy primitives and ϵ's. In the figure, $?a$ denotes reception of message a, and $!b$ denotes transmission of message b. In Figure 7, the process A selects only one service at the state after reception of message c (expressed by ●). Similarly, the protocol specifications of Processes C and D are derived.

Figure 8 shows a sequence chart representing the protocol specification derived from CW and CFV services for which a feature interaction occurs. In this figure, black thick lines represent the sequence in the case that CW service is selected, and gray thick lines represent the sequence in the case of CFV service. At a state denoted by ● one of these services are dynamically selected during their execution.

6. Evaluation of Proposed Method

We consider feature interactions between six telephone services: Three Way Calling (TWC), Call Forwarding (CF), Call Waiting (CW), Denied Origination (DO), Direct Connect (DC), Denied Termination (DT) services, which are described in [14]. Circles in Figure 9 show the fact that feature interactions occur for each pair of two services, which ATR has detected by experiments on reachability analysis of system states[13].

In particular, four types of feature interactions occur for CW and TWC services. We explain these four cases using event sequences of CW and TWC services in Figure 10. Feature interactions occur (1) when events flash(A) are executed at a system state including System State CW2 and System State TWC1, (2) when events flash(A) are executed at a system state including System State CW3 and System State TWC1, (3) when events flash(A) are executed at a system state including System State CW2 and System State TWC5, and (4) when events flash(A) are executed at a system state including System State CW2 and TWC3.

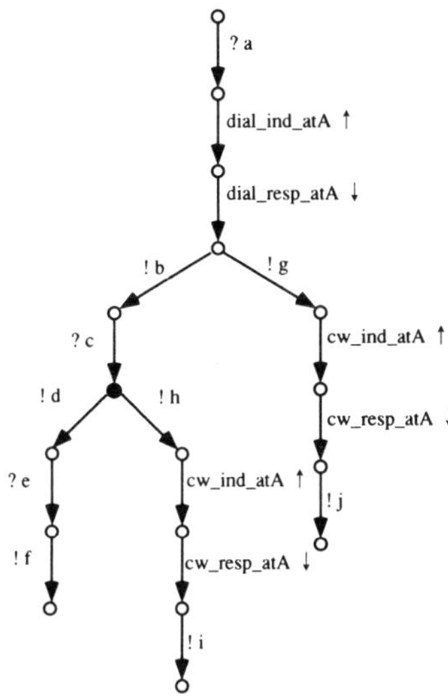

Figure 7: Protocol Specification for Process A

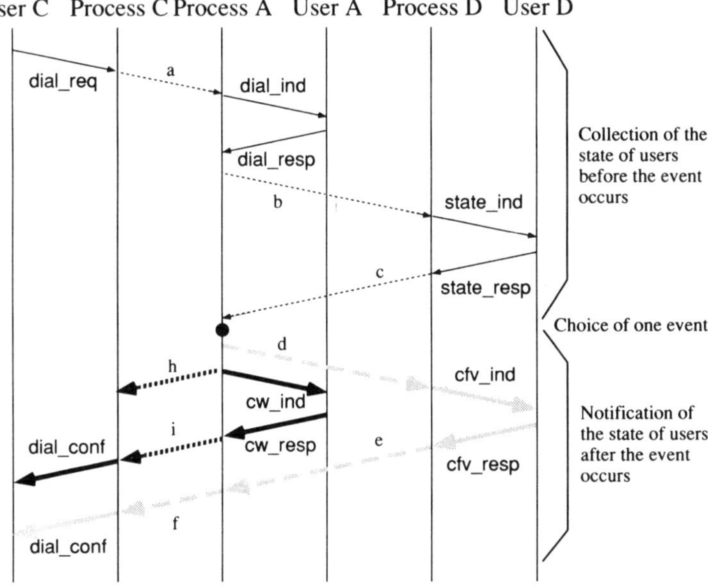

Figure 8: Sequence Chart Representing Protocol Specification Generated by Protocol Synthesis

Figures 11-17 show service rules for which a feature interaction occurs and resultant sequence charts which are automatically derived by the proposed method. The following observation is obtained from these results. States of users at the current state of the service rule can change during collection of the states because users can independently execute events with each other. In other words, even when predicates are satisfied before their collection, there may exist some predicates, which are not satisfied when the states are collected. For example, as shown in Figure 11, user C can make ringback(C,A) and r-path(C,A) false during user A collects the state of user C, and as shown in Figure 13, user D can make ringback(D,A) and r-path(D,A) false during user A collects the state of user D. In the proposed method, since an event for which a feature interaction occurs is selected according to not only (Criterion 1) but also (Criterion 2), execution of the event is always assured after their collection. Compared with the method in the service specification, the proposed method in the protocol specification thus enables flexible selection of events when feature interactions occur.

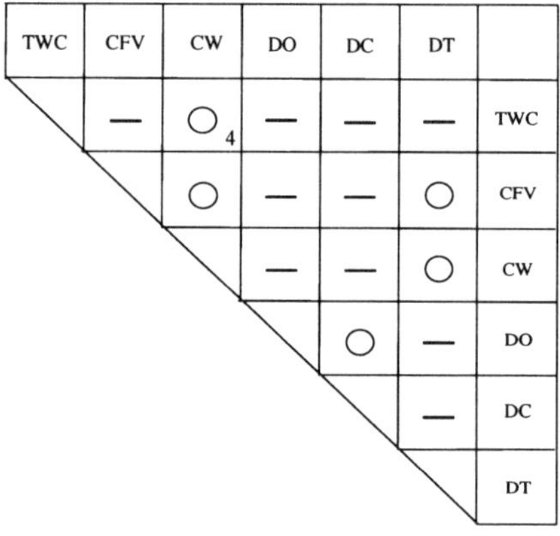

Figure 9: Feature Interaction between Six Services

7. Conclusion

This paper has proposed a dynamic method for resolution of the feature interaction problem, which is based on the protocol synthesis technique. The feature interaction can be interpreted as a nondeterministic behavior of executable events in the service specification. Since each event in the service specification is translated into a sequence of primitives in the protocol specification, the feature interaction is regarded as a non-deterministic execution of primitive sequences and transmission of messages. The first half of one sequence of primitives and messages includes that of the other, but the latter half is different with each other. Thus, in the proposed method, the feature interaction problem can be resolved by inserting, to the protocol specification, conditional branch functions for determining which primitive sequence is selected.

By applying the proposed method to feature interactions between several telephone services, it is confirmed that feature interactions detected in the service specification can be resolved in the protocol specfication which is automatically derived by the proposed method and that execution of an event for which a feature interaction occurs is assured by dynamic selection of the event based on the collected states of users.

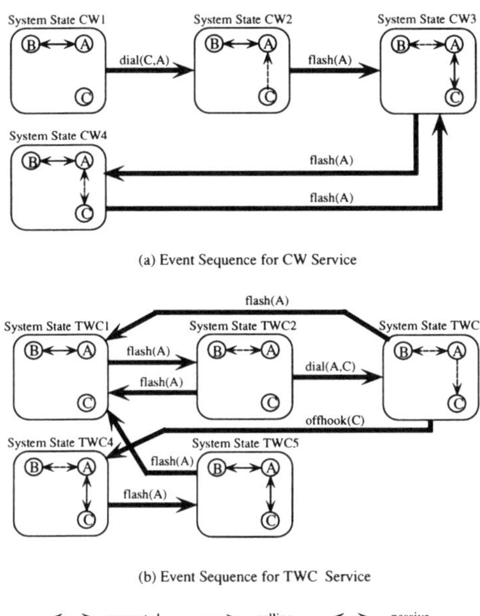

Figure 10: Event Sequences for CW and TWC services

Service Rules

CW   path(A,B) , ringback(C,A) , r-path(C,A) , cw-ringing(A,C)
flash(A) :
cw(A) , path(A,C) , path(C,A) , path-passive(A,B) .

TWC   path(A,B) , m-3wc(A)
flash(A) :
path-passive(A,B) , dial-tone(A) , 3wc(A) , m-3wc(A) .

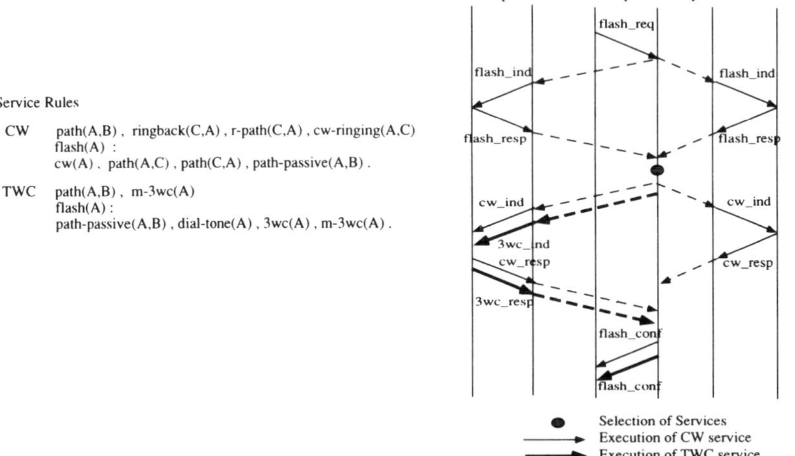

Figure 11: flash(A) executed at System States CW2 and TWC1

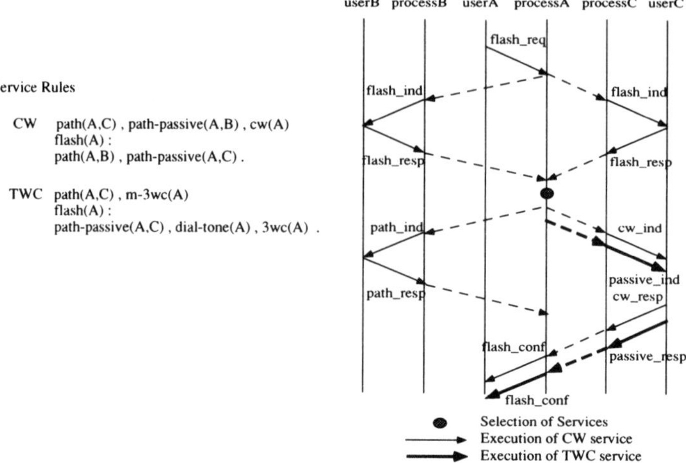

Figure 12: flash(A) executed at System States CW3 and TWC1

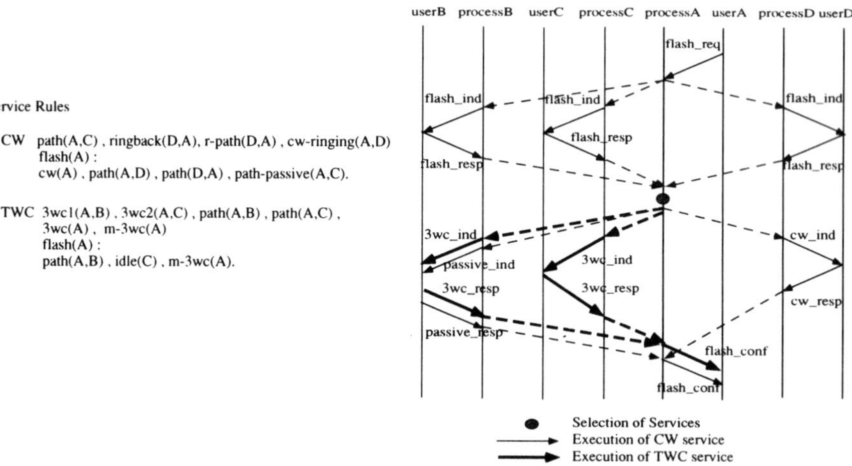

Figure 13: flash(A) executed at System States CW2 and TWC5

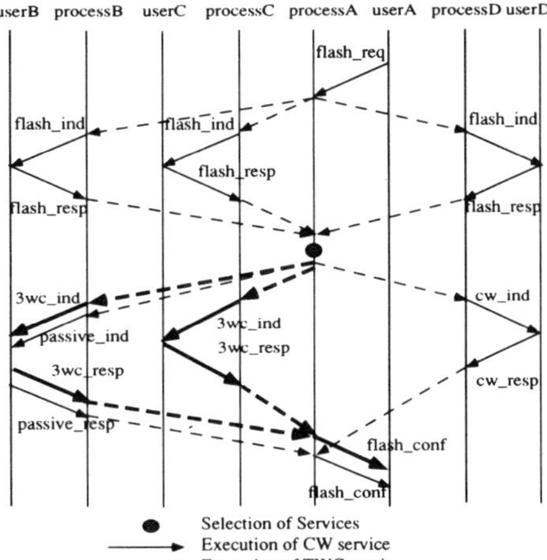

Service Rules

CW path(A,B) , ringback(D,A),r-path(D,A),cw-ringing(A,D)
      flash(A) :
      cw(A) , path(A,D) , path(D,A) , path-passive(A,B).

TWC 3wc1(A,B) , 3wc2(A,C) , path(A,B) , Calling(A,C) ,
      3wc(A) , cond:m-3wc(A)
      flash(A) :
      path(A,B) , idle(C).

Figure 14: flash(A) executed at System States CW2 and TWC3

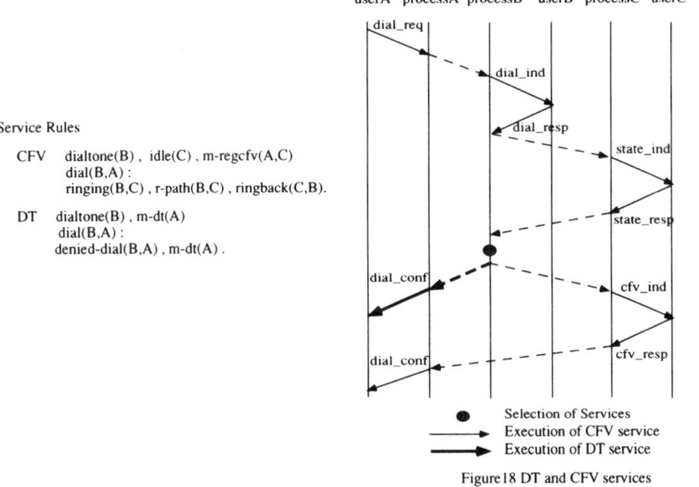

Service Rules

CFV   dialtone(B) , idle(C) , m-regcfv(A,C)
        dial(B,A) :
        ringing(B,C) , r-path(B,C) , ringback(C,B).

DT   dialtone(B) , m-dt(A)
        dial(B,A) :
        denied-dial(B,A) , m-dt(A) .

Figure 15: DT and CFV Services

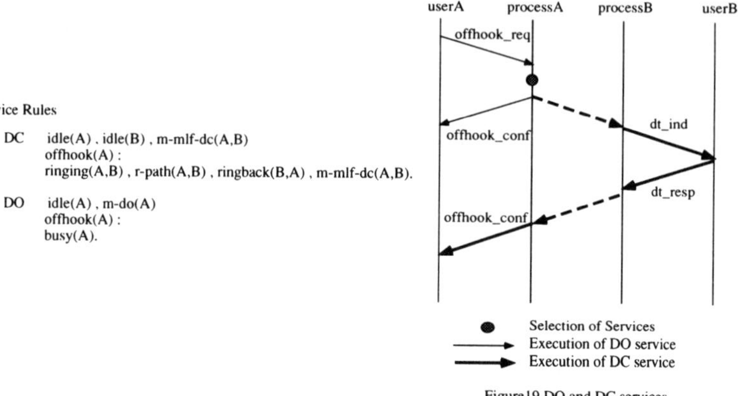

Service Rules

DC   idle(A) , idle(B) , m-mlf-dc(A,B)
      offhook(A) :
      ringing(A,B) , r-path(A,B) , ringback(B,A) , m-mlf-dc(A,B).

DO   idle(A) , m-do(A)
      offhook(A) :
      busy(A).

Figure 16: DO and DC Services

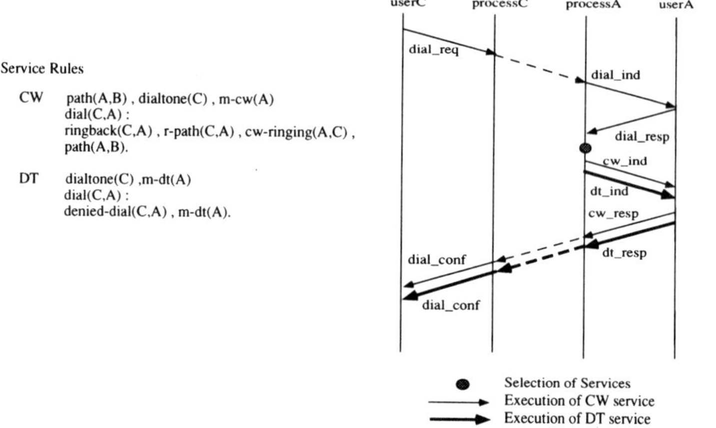

Service Rules

CW   path(A,B) , dialtone(C) , m-cw(A)
      dial(C,A) :
      ringback(C,A) , r-path(C,A) , cw-ringing(A,C) ,
      path(A,B).

DT   dialtone(C) ,m-dt(A)
      dial(C,A) :
      denied-dial(C,A) , m-dt(A).

Figure 17: CW and DT Services

References

[1] Bellcore, "LSSGR Features Common to Residence and Business Customers I, II, III," Issue 2, July 1987.

[2] H.Bussey, S.Minzer, P.Mouchtaris, S.L.Moyer, and T.F.Porta "EXPANSE software for distributed call and connection control," International Journal of Communication Systems, Vol.7, No.2, 1994.

[3] P.M.Chu and M.T.Liu,"Protocol synthesis in a state transition model," Proc. COMPSAC'88, pp.505-512, Oct.1988.

[4] N.Griffeth and Y.-J.Lin(eds.), Special issue on managing feature interactions in telecommunications systems, IEEE COMPUTER, Vol.26, No.8, Aug.1993.

[5] Y.Harada, Y.Hirakawa, T.Takenaka and N.Terashima,"A conflict detection support method for telecommunication service descriptions," IEICE Trans. Commun, E75-B(10), pp.986-997, Oct.1992.

[6] Y.Hirakawa and T.Takenaka,"Tecommunication service description using state transition rules," Proc. Sixth Int'l Workshop on Software Specification and Design, pp.140-147, Oct.1991.

[7] H.Igarashi, Y.Kakuda and T.Kikuno,"Synthesis of protocol specifications for design of responsive protocols," IEICE Trans. on Information and Systems, E-76-D(11), pp.1375-1385, Nov.1993.

[8] Y.Kakuda, H.Asada and T.Kikuno,"Application of protocol synthesis technique to resolution of the service interaction problem," Proc. Seventh Int'l. Conference on Formal Description Techniques(FORTE'94), pp.369-372, Oct.1994.

[9] Y.Kakuda, H.Igarashi and T.Kikuno,"Automated synthesis of protocol specifications with message collisions and verification of timeliness," Proc. of Second Int'l. Conference on Network Protocols(ICNP'94), pp.143-150, Oct.1994.

[10] Y.Kakuda, H.Asada and T.Kikuno,"Protocol synthesis technique and its application to the feature interaction problem," Proc. ATR International Workshop on Communications Software Engineering, pp.96-110, Oct.1994.

[11] Y.Kakuda, M.Nakamura and T.Kikuno,"Automated synthesis of protocol specifications from service specifications with parallelly executable multiple primitives," IEICE Trans. on Fundamentals of Electronics, Communications and Computer Sciences, pp.1634-1645, Oct.1994.

[12] Y.-J.Lin and N.Griffeth(eds.), Special issue on managing feature interactions in telecommunications systems, IEEE COMMUNICATIONS MAGAZINE, Vol.31, No.8, Aug.1993.

[13] M.T.Liu,"Protocol engineering," Advances in Computers, 29, pp.79-195, Academic, 1989.

[14] T.Ohta and Y. Harada,"Classification, detection and resolution of service interactions," Feature Interactions in Telecommunications Systems, pp.60-72, IOS Press, 1994.

[15] T.F.Porta and M.Veeraraghavan, "Design and implementation of a distributed call processing architecture," Proc. of ICC'94, 1994.

[16] K.Saleh,"Automatic synthesis of protocol specifications from service specifications," Proc. Int'l. Phoenix Conf. on Computers and Communications(IPCCC'91), pp.615-621, Mar.1991.

[17] K.Saleh and R.L.Probert,"Synthesis of communication protocols: Survey and assessment," IEEE Trans. Comput., 40, 4, pp.468-475, Apr.1991.

[18] A.Takura and T.Ohta,"Stepwise telecommunication software generation from service specifications in state transition model," Proc. of Second Int'l. Conference on Network Protocols(ICNP'94), pp.135-142, Oct.1994.

[19] M.Wakabayashi, Y. Kawarasaki and T.Ohta, "An evaluation of service interaction detection method at specification description," Proc. of the 1994 IEICE Fall Conference, B-542, Sept. 1994, in Japanese.

Automatic Reconfiguration for Runtime Feature-Interaction Resolution in an Object-Oriented Environment

Petre DINI, Gregor v. BOCHMANN

Departement of Computer Science and Operational Research
University of Montreal, CP 6128, Succursale CENTRE-VILLE
Montreal, H3C 3J7, Canada

Abstract. In this paper, we consider that the following three distinct factors could ameliorate the problem of feature interactions in object-oriented environments namely, (1) a well formal definition of the domain-environment boundary, (2) a well description of reconfiguration state of managed objects representing real system components, and (3) adapted reconfiguration policies to avoid, detect, and diagnose feature interactions across a domain-environment boundary in a given reconfiguration state. Consequently, we present an approach to cope with feature interactions within object-oriented environments in the case of automatic reconfiguration from the survivability perspective. Previous ideas are presented across three examples of feature interactions concerning the existential boundary, the test boundary, and stand-by substitutability aspects.

Key words: *object-oriented, feature interactions, automatic reconfiguration*

1. Introduction

Several complex management problems arise within open distributed systems due to their *heterogeneity* with respect to the hardware, software, and architecture, their *distributivity* concerning different aspects of communications, synchronization, localization, and migration, or to their *openness* insuring the transparency of structure and behavior changes.

Complex system functionality implies specific various *management activities* with respect to the resource allocation, the system dynamic evolution, and the evaluation of this evolution in order to offer the quality of service desired by the users. These tasks are usually performed either by mono-processor operating systems, or by highly independent cooperating operating systems in mono-processor systems, or in cooperating centralized systems, respectively. Since dynamic changes or unexpected behaviors due to users, or to resource disponibilities are often possible within open distributed systems, management aspects became more complex. Gradually, these aspects have been separated as management tasks fulfilled by special management applications.

This work was partially funded by the Ministry of Industry, Commerce, Science and Technology, Quebec, under IGLOO project organized by the Computer Research Institute of Montreal, and by a grant from the Canadian Institute for Telecommunication Research (CITR) under the Networks of Centres of Excellence Program of the Canadian Government.

The authors' address is Computer Research Institute of Montreal, 1801, McGill College street, Suite 800, Montreal, Qc, H3A 2N4, CANADA, phone: (514)398-1608, fax: (514)398-1244, e-mails: {dini, bochmann}@{crim.ca, iro.umontreal.ca}

These applications use particular information data, special protocols, and distinct policies and offer services within several distinct areas such as the system configuration, faults aspects, etc. *Automatic management policies* in large distributed systems or in evolving systems are strongly dependent on *services interactions* and *feature interactions*. Feature interactions are more visible in *automatic reconfiguration* decisions implying different kinds of hardware and software components. Furthermore, particular aspects appear in object-oriented environments due to the object interaction and the lack of formal description of *interobject behavior.*

In this paper, we consider that the following three distinct factors could ameliorate the problem of feature interactions in object-oriented environments namely, (1) a well formal definition of the domain-environment boundary, (2) a well description of reconfiguration state of managed objects representing real system components, and (3) adapted reconfiguration policies to avoid, detect, and diagnose feature interactions across a domain-environment boundary in a given reconfiguration state. Consequently, we present an approach to cope with feature interactions within object-oriented environments in the case of automatic reconfiguration from the survivability perspective. Before, we present relevant definitions in order to smoothly pass from the object-oriented context to the feature interactions context.

1.1. Quality, evolution, and survivability perspectives

Several concerns lead to requests for reconfiguration of systems such as *system evolution, fault tolerance enhancement, system stability,* or *QoS enhancement.* We classify the reconfiguration management actions from three distinct perspectives namely, the *survivability perspective,* the *quality perspective* and the *evolution perspective.* The *evolution perspective* is a problem often encountered in open systems. Components, component types, services are constantly being added or removed. Furthermore, their parameters or relations vary continuously. The quality perspective is related to the enhancement of quality of service, an optimal configuration, or a good system stability. The *survivability perspective* is a difficult issue even for the simpler centralized systems where there is a single manager (in fact, the operating system). In open distributed systems there are two opposite aspects. First, the openness increases the possibility of incompatibility between components, while the distribution raises problems of global state and of correlations between widespread domain managers. On the other hand, there is a larger possibility to redistribute tasks and to use various service types, possibly redundant.

1. 2. Basic Terminology

Object-oriented issues

We define first, the relevant object-oriented terms related to the feature interactions. Each software or hardware system entity is modelled as an object which encapsulates data as *attribute values* and the behavior as methods implementing its own *operations.* The *operation signature* are properties that must be satisfied for each execution of the operation (independent of the execution history of the object), such as the following: (1) the number of arguments, (2) the number of results, and (3) syntactic typing constraints namely, for each argument and result, the type to which the effective argument or result object should belong. The list of all operations provided by an object together with their signature defines the *object interface.* Different scopes can determine an object to have *multiple interfaces* such as functional interfaces, referring to the functional role of the object within a DCS, or management interfaces dedicated to management goals. The *object signature* is defined as the entire set of object operations, together with ordering constraints (accepted or refusal sets), or error handling issues.

Interobject behavior refers to the object cooperation to perform a common target. This cooperation determines a specific period in the object life-cycle, where the behavior of one object is dependent or in accordance with the behavior of other objects. In a large vision, interobject behavior focuses on *interactions* and *dependencies* between the objects of a system, i.e., for each interobject behavior an object interacts with its environment, or between the objects of a sub-system and its environment. We consider the interobject behavior as constraints imposed to objects. Constraints can be described as predicate-conditions on individual behavior [4]. Interobject behavior concerns also the interaction between a system and its environment, viewed as aggregated objects. A special case of interobject behavior is the *environmental behavior*, where objects cooperate with the physical environment. Environment becomes important because interobject interfaces will be implemented differently, depending whether the part of the system is realized in software, hardware, mechanically, or through interaction with the user. Consequently, the need to clearly define the environment and rules applying when objects of a system interact with objects of the system's environment becomes of relevance.

Feature interaction issues
We translate now the object-oriented terms into those specific to feature interactions. The operation signature defines the *operation features*. A *service* is a set of operations offered by an object. Since an object can be composed by aggregation, this definition contains that given in [1], i.e., "a service is a meaningful set of capabilities provided by or over a network to its different players, including end users, network providers, and service providers". *Service features* represent operation features of services contained in the service, and other features imposed by the operation ordering and error handling cases. Since operation features are included in services features as a particular case where the service has only one operation, we use hereafter the term service features for both situations. *Interobject behavior* corresponds to either *service interaction*, or *service interference*. *Service interactions* are determined by *intended interactions* when a service is affected because of another service, e.g., the capability enhancement performed by the designer. However, there are many *unintended interactions,* called *service interferences* by Mierop et al [1].

The e*nvironmental behavior* is considered as a *service interaction* or *inference aspects*, where the "interaction is an occurrence in which the behavior of an entity within a specific environment is affected by the behavior of another entity within the same environment" [5].

The distinction between a *service* and a *feature* may depend on the context of interpretation. For example, an automatic reconfiguration policy within a system is a service from the system point of view. However, from the management operator point of view the automation is a system feature.

We define in an object-oriented environment, a *feature* as either a part of the operation, service or object signature, or properties of object interactions, expressed as dependency constraints between participating objects. This dependency may underlines some aspects, such as business, temporal, or implementation. Our definition includes that presented by Velthuijen which status that "a feature in a telecommunications system is a package of functionality incrementally added to a service to enhance or modify it" [2]. This corresponds to the evolution perspective and quality perspective, but do not cover the survivability perspective.

Consequently, *feature interactions* names the process resulting between cooperating objects which offer to each other their own services. We consider that the *feature interaction problem* is not only as presented in [2], i.e., "the detection, prevention, and resolution of undesired feature interactions", but also, the validation between features of older and newer component in the case of substitutability, the optimization of existing desired feature interactions, or the automatic adaptation to desired changes of feature interactions.

2. Motivation

The evolution of management applications as particular telecommunication software becomes of a major importance. Automation of reconfiguration policies requires new approaches taking into account runtime features [3], negotiation aspects [6][8], cooperation relations [9], or system learning [7]. We quote below recent opinions which emphasize the need of automation.

> - "..when something *unanticipated happens* in the environment, system entities and their relationships may need to be reorganized in order to accommodate new requirements...." or "...it is no longer sufficient to provide a model of communication.... systems must assume a more positive emphasis, such as *providing explanations* or *suggesting alternative possibilities*, placing emphasis on system *learning*". The situation reduces to two principles: (i) how to cope with *events that fall outside solutions* offered within the current conceptual framework, and (ii) how to ensure the system will be able to *make use of such information for the future* [7].

> - ".. a desirable feature of network devices if, on being plugged into the network, they could query each other's configurations, query each other's service capabilities, query each other's supported protocols, negotiate an acceptable protocol suite among themselves, negotiate options and parameters associated with the protocol suite, and automatically configure themselves for service - all with minimal or zero human intervention. This capability is called *autoconfiguration*." [6].

Among other problems, feature interactions are twofold vital for decisions using automatic policies for reconfiguration namely, (1) feature interactions determine which kinds of policies must be applied, and (2) feature interactions validate the results of such policies. Furthermore, the fact that ISO/OSI [10] and ISO/ODP [11] have taken the object-oriented approach to abstractly represent system components, and that such of object-oriented representation has been partially implemented in the ongoing IGLOO project [12], our proposal is described in the same approach.

2.1. Related work

Different approaches of formalizing relevant information to model feature interactions exist. An approach similar to finite state machine, called concurrent communicating automata is analyzed in [13]. Cameron and Velthuijsen reports other approaches close to the telecommunications [3]. For instance, Vanslembrouck&Verdonk's approach is based on "specifying the connections between parties in a communication session, rather than the sequences of states that define the call processing logic", whereas the Bellcore's Touring machine architecture is based on objects and their interfaces. They propose a taxonomy of the feature definition considering the network configuration view as one of five ways to view the feature interaction problem. Mierop et al present an architecture where service specification and resource specification are two separate frameworks [1]. They present service interferences into an ISO/ODP perspective and analyze trading aspects for the callee-called problem.

However, the automation is not directly expressed. Even further, "the traditional manual approach is no longer adequate" [3], whereas the current techniques to automatically detect feature interactions are in study. We consider that specific views on managed objects representing system components must be considered for automatic reconfiguration in order to avoid undesired feature interactions. Also, the boundary between the system solving a problem and its environment must be well-defined in order to capture undesired feature interactions.

2.2. Problem

We study in this paper feature interactions raised in automatic reconfiguration management. We deal first with the question where the automatic manager must verify first potential feature interactions in order to consequently react. In that sense, we define the notion of domain-environment boundary, which is useful in feature interactions raised by the domain substitutability. Second, we propose object qualifications to automatically recognize potential feature interactions by a specific object state attribute called reconfiguration state.

The remainder of this paper is structured as follows. Section 3 presents the notion of system-environment boundary where features interactions are detected and assumptions at this boundary. The configuration state attribute and the diagram of configuration state changes are also defined. Section 4 underline the previous ideas across three examples of feature interactions concerning the existential boundary, the test boundary, and stand-by substitutability aspects. Section 5 concludes on our proposals and presents the future work.

3. Proposal

3.1. System-environment boundary

Substitutability problem as feature interactions
When a system is in operation, it may require changes including but not limited to the substitution of a service by another service offering better performances (perfective maintenance) or the addition of a new service (adaptive maintenance). Within an object-oriented system, its services are realized by objects or by collections of interacting objects (known as subsystems). Parnas while studying the architecture of software systems found that *the connections between modules are the assumptions that the module makes about other modules* [14]. As a consequence, any change to a module may modify the assumptions it satisfies with respect to the remaining modules it is connected to. This means that when a service is realized by an object (subsystem), a change to this service has some impacts on the objects (subsystems) connected to this object. Further, this may in turn have impacts on the new service itself, because when it modifies the assumptions made by other services, this action may alter these services so that the hypotheses this service assumes are no longer valid and in turn provoke a reaction upon this service. Therefore, when considering the substitution of a service, ultimately an object (subsystem), we may ask ourselves if this new service, ultimately the new object (subsystem) *will properly behave in its target environment.* On the other hand, when we add a new service, ultimately a new object (subsystem), we

may wonder if *the environment of this service still maintains (subsumes) the assumptions relative to the situation before this adjunction.* These are important questions for the maintenance and evolution of systems. Our assertion is the assumptions made about the environment of objects, subsystems and systems must be known and understood prior to any modification of the system. This is not something new, but put in the context of object-oriented systems, and of the automation of the reconfiguration management, testing and maintenance of distributed systems it becomes a cornerstone of system modeling in our days.

Assumptions

Referring to the object-oriented development process, we interpret in the following the assumptions. We distinguish three levels of representation for the environment namely, the *environment of the system*, the *environment of each subsystem within this system*, and the *environment of each object forming this system*. At the analysis phase, we define the *environment of the system*. The constraints we place on the objects forming the system reflect the assumptions we make about its environment. The analysis model consists of a set of interacting subsystems and objects. These objects and subsystems make assumptions about the objects and subsystems to which they are connected. This creates for the analysis model two kinds of assumptions, the external assumptions (system's environmental assumptions) and the internal assumptions about objects and subsystems forming this system.

Since *object*, *sub-system*, and *system* are substitutable notions with respect to the system-environment interaction, we call all these three terms *domain* (D). A *system* is initially designed starting from the user's needs. Environment and system components interact across cooperation relations. The environment of a domain is the set of objects with which the domain components interact at a certain time in given conditions. The domain-environment *boundary* is the set of all cooperation relations whose interacting objects belong separately to domain and environment.

Environment (E) component set of an object comprises the following objects: (1) the objects from which this object has at least one reference (static or dynamic, i.e. through an attribute or as an input parameter of the object's method or also an output parameter form a operation call to another object), and (2) the object which hold a reference to this object using the mechanism elaborated in (1). In the following representations we formally summarize the boundary (B) for an object. It extend the informal definition of the interface used by the Fusion method [19]. If two objects X and Y have a cooperation relation $\mathcal{R}$, all their interactions occur through this relation, i.e., X$\mathcal{R}$Y (read X related to Y). In the following, we define an interaction model for an object X, at the instance of time t.

$$D(t) = X;$$
$$E(X, t) = \{O \mid X\mathcal{R}O \lor O\mathcal{R}X\}, \text{ the X's environment;}$$
$$B(X, t) = \{\mathcal{R} \mid X\mathcal{R}O \lor O\mathcal{R}X\}, \text{ the X's boundary.}$$

Extensions to multi-objects domains and multi-domains of this model are presented in [15]. Based on this model distinct kinds of environments have been proposed. According to them, there are distinct kinds of feature interactions across different types of boundaries: existential, functional, test, and management. Two kinds of assumptions concern feature interactions, i.e., assumptions about the environment objects, and assumptions about boundary relations. Since the conditions offered by the environment are by no means static and they may change with time, some assumptions are true at a certain time and false at other points of time. The assumption about the environment of a domain D at a certain time t is a formula φ in some logic on the objects forming E(D,t). It is denoted by the formula $\varphi(E(D, t))$. Analogously, the assumptions about the domain-environment boundary are denoted by a formula $\psi(B(D, t))$.

3.2. Configuration state

Real resources of distributed systems are represented in MIBs (Management Information Base) by *managed objects* which have several attributes describing the *state* of the associated distributed system resource. Their value space is standardized as follows: *operationalState* (*enable, disable*), *usageState* (*idle, active, busy*), and *administrativeState* (*locked, shutting down, unlocked*) [18]. A specific feature called *configurationState* has been introduced in [17] to precise the real perspective from the reconfiguration point of view. Configuration state does not affect the managed object's ability to respond to management operations. *Configuration state* is imposed through the *configuration services* and describes the *permission* to use or *prohibition* against using the resource, with respect to *configuration actions* such as state change or migration of a component, and create or destroy cooperation relations belonging to the domain-environment boundary.

Some classes of managed objects exhibit only a part of proposed configurationState values. For example, a hardware link needs automation for neither insertion, nor initialization. On the other hand, the *configurationState* of an abstraction of a software application can take any configuration state value. The configuration state is read/write in nature. The new attribute of reconfiguration state is represented by *reconfigurationState* (*inserted, initial-config, allowable, waiting, in-active-tests, in-passive-tests, in-changes, inactive, isolated*), where each value represents distinct configuration phases of a system component.

`inserted`: A new system component is added to a management domain. If the component type is new, the manager must update the existing type repository, verify the conformance of type, and update the MIB with a new component. Otherwise, only the MIB must be updated.

`initial-config`: A new resource commonly presents its *requestedEnvironment* [9] feature representing the allocation view (one or more requested cooperation relations, as a precondition to offer its services). This value describes the state where all requested relations are build. Some attributes are initialized.

`allowable`: In this state no reconfiguration actions are performed or predicted. This state permits to consider existing standardized state values for other state attributes.

`waiting`: Because of external conditions or of the usage, operational or administrative attribute value, a *configuration service* has been requested.

`in-active-tests`: Relations of concerned objects are updated with those specific to tests. This state value represents a cold testing, where the component does not offer its services, and administratively it is locked. all the cooperation relations are "sleeping", only the allocation relations are "fired". In this way a resource is isolated from the system and tested without disturbing the system state [16].

in-passive-tests:This state represents a specialization of the *allowable* state. Passive tests control different aspects of component behavior without interrupting its normal services. This state is not dependent of usage state value.

in-changes:Attribute values, component relations are changing. For example, a preparation for migration to other domain, for removing, or for out-of-work passes necessarily through this state.

inactive:The component is not available because of reconfiguration reasons. In this state all previous allocation relations hold, but their state is "sleeping". Correspondent resources are free to entirely offer their services to other customers. For example, a *cold stand-by* resource has this configuration state value.

isolated:The component is not available for reconfiguration goals, its relations are *destroyed*. For a new use, if the component does not change its management domain, a new initialization is required. Otherwise, the component must be inserted first.

Configuration services must ensure the *transitions* shown in Figure 1 by specific *configuration commands*. In Section 4.2 we present the transition from allowable to in-passive-tests, as an example.

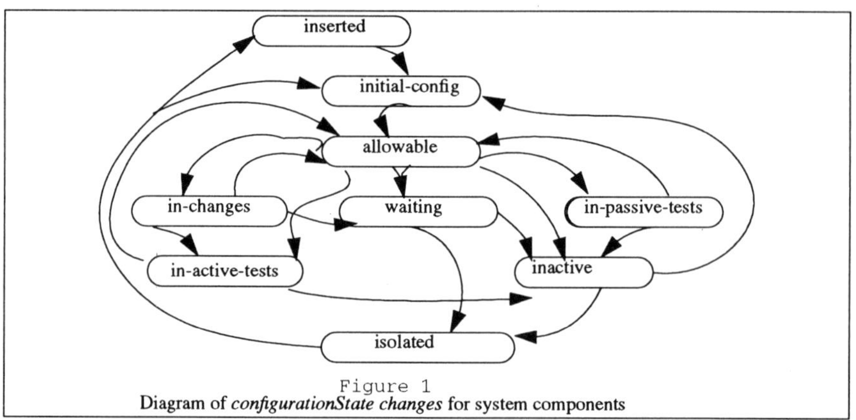

Figure 1
Diagram of *configurationState changes* for system components

4. Application of our proposal

4.1. Feature interactions at the existential boundary

Let us exemplify the concept of environment and assumption about the environment. Consider a system $D(t_1) = \{X, Y\}$, its environment $E(D, t_1) = \{Z\}$, and their frontier $F(D, t_1) = \{\mathcal{R}_1,$

$\mathcal{R}_2 \mid X\mathcal{R}_1 Z \wedge Y\mathcal{R}_2 Z\}$. We also assume that interacting with an object means that the object may trigger some operation at the cooperating object. Let us consider that the $\mathcal{R}_2$ relation belongs to an existential frontier. At the time t_2, the object Y deletes the object Z and removes its reference to Z from its state's attributes. This is illustrated by the following.

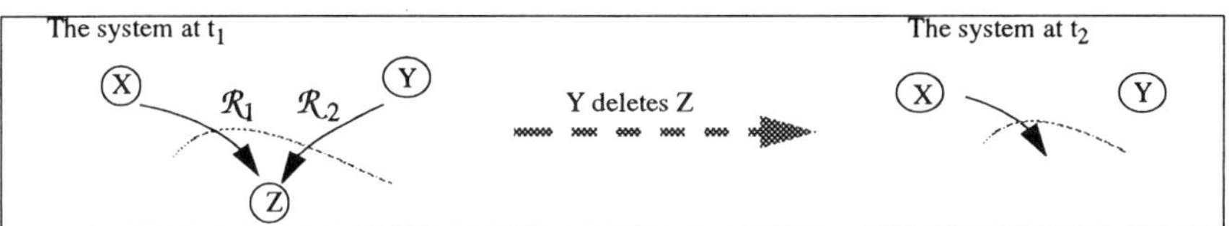

We restrict (without loss of the generality) the assumptions about the environment relative to this situation to the existence within the environment of the objects another object holds a reference to. When the environment is empty, the assumption about this environment is true at all the time the environment remains empty. Therefore, for any environment E(D,t), the related assumption is of the form $\varphi(E(D, t)) \equiv \exists \alpha \mid \alpha \in E(D, t)$ *if* $E(D, t) \neq \varnothing$, *or true if* $E(D, t) = \varnothing$.

It means that the object may have a reference to another object, but this reference does not necessarily reference an existing object. *The assumption about the environment consists in supposing the existence of this object.*

The system at time t_1 can be described by the following facts:

$$D_1(t_1) = \{X\} \qquad E(D_1, t_1) = \{Z\}; \qquad B(D_1, t_1) = \{\mathcal{R}_1\}.$$
$$D_2(t_2) = \{Y\} \qquad E(D_2, t_1) = \{Z\}; \qquad B(D_2, t_2) = \{\mathcal{R}_2\}.$$
$$D = D_1 \cup D_2 \qquad E(D_1 \cup D_2, t_1) = \{Z\} \qquad B(D, t_1) = \{\mathcal{R}_1, \mathcal{R}_2\}.$$

and the assumptions are:

$\varphi(E(X, t_1)) \equiv \exists Z.\ Z \in E(X, t_1)$ because at t_1, E(X, t_1) = {Z}

$\varphi(E(Y, t_1)) \equiv \exists Z.\ Z \in E(Y, t_1)$ because at t_1, E(Y, t_1) = {Z}

$\varphi(E(Z, t1)) \equiv \exists X.\ X \in E(Z, t_1) \wedge \exists Y.\ Y \in E(Z, t_1)$ because at t_1, E(Z, t_1) = {X, Y},

whereas at the time t_2

$E(X, t_2) = \varnothing$

$E(Y, t_2) = \varnothing$

and the assumptions are

$\varphi(E(X, t_2)) \equiv \exists Z.\ Z \in E(X, t_2)$ which is false because Z does not exist at the time t_2.

$\varphi(E(Y, t_2)) \equiv$ true because at the time t_2, E(Y, t_2) = $\varnothing$.

4.2. Feature interactions using reconfiguration state

The following sequence present feature interactions according to our extended definition presented in Section 1.2, using the reconfiguration state and the domain-environment boundary previously defined. Interactions refer here to the transition from the configuration state *allowable* to the configuration state *in-passive-tests*.

- component for tests:
    - administrativeState --> unlocked | shutting-down
    - operationalState = = enable
    - configurationState --> in-passive-tests
- testing component:
    - administrativeState --> enable
    - operationalState --> enable
    - configurationState --> allowable

- actions of automatic reconfiguration components:
    - identify the appropriate testing component at the test boundary;
    - connect it to requested environments (components which offer services to it, such test monitor, test scheduler, or test synchronizers) at the existential boundary and the functional boundary;
    - connect it to the component for test, across the functional boundary;
    - connect the test analyzer, across the functional boundary;
    - filter decisions of the test analyzers;
    - send relevant decisions to the component manager, across the management boundary.

- actions of the manager level:
    - passively monitor the observation points, across the management boundary;
    - interpret the behavior of the component in tests;
    - correct the configuration if the test results do not guarantee the safety of the component in tests, across the management boundary and the configuration state.

4.3. Feature interactions in reconfiguration of stand-by components

In this section, we present feature interactions concerning a particular aspect of substitutability, i.e., substitutability of network hardware components. Some hardware features refer to the quality of service (QoS) offered to potential customer, e.g., processing speed, error rate, capacity, overload tolerance. Some component features are temporarily stable, e.g., the number of free channels on a path, a maximum delay replay, whereas particular features are dependent on other components. Among the latter reconfiguration features making easy the automatic reconfiguration of hardware system components are (1) tactical QoS provision for a short time, e.g., to save critical values in case of a failure or to ensure a graceful degradation of the optimal QoS, and (2) mobility, i.e., mobile station can connect itself to another server-station to get a better QoS or to save a critical logical link.

There are three kinds of feature interactions which may favor the automatic reconfiguration. For each of them, we subsequently present feature interactions taking place at different types of boundaries and implying different reconfigurationState values.

- there is an identical stand-by entity either connected or disconnected
    - in the connected case the commutation will be directly done

  · reconfigurationState = allowable
  · reconfiguration actions: >at the functional boundary
 • in the disconnected case some delay will be unavoidable
  · reconfigurationState = initial-config
  · reconfiguration actions: >at the existential boundary (connections for
          initialization)
          >at the functional boundary (connections as server);

• there is a similar stand-by entity offering a degradation of QoS
    • in the connected case the commutation will be directly done
  · reconfigurationState = allowable
  · reconfiguration actions: > at the management boundary (validate degradation)
          > at the functional boundary

 • in the disconnected case some delay will be unavoidable
  · reconfigurationState = initial-config
  · reconfiguration actions: > at the management boundary (validate degradation)
          >at the existential boundary (connections for
          initialization)
          >at the functional boundary (connections as server);

• there is a predefined priority list of possible substitutes
  · reconfiguration actions: > at the test boundary
          > at the management boundary
          > at the functional boundary

5. Conclusion

In this paper, we have presented considerations and issues on feature interactions concerning the management aspects within an object-oriented environment. First, quality, evolution, and survivability perspectives have been defined. Therefore, we have translated object-oriented issues to feature interaction issues, extending feature interactions to validation, optimization, and automatic reconfiguration aspects. Especially, we have studied feature interactions raised in automatic reconfiguration management. We have defined the system-environment boundary in order to answer the question where the automatic manager must verify first potential feature interactions in order to consequently react. Second, we have proposed means to automatically recognize potential feature interactions by a specific object state attribute called reconfiguration state. Based on these proposals and using some assumptions types we have presented three examples of feature interactions concerning the existential boundary, the test boundary, and stand-by substitutability aspects.

Ongoing work is being done on the classification of feature interactions at the system-environment boundary and the design of appropriate automatic reconfiguration policies [9].

References:

[1] John Mierop, Stefan Tax, Ronald Janmaat, "Service Interaction in an Object-Oriented Environment", *IEEE Communication Magazine*, August 1993, pp. 46-51

[2] Hugo Velthuijsen, "Distributed Artificial Intelligence for Runtime Feature-Interaction Resolution", *IEEE Computer*, August 1993, pp. 48-55

[3] E. Jane Cameron, Hugo Velthuijsen, "Feature Interactions in Telecommunications Systems", *IEEE Communication Magazine*, August 1993, pp. 18-23

[4] Gregor v. Bochmann, Petre Dini, Michel Barbeau, Piero Colagrosso, Rudolf Keller, Dunia Ramazani, Stéphane Somé, "Common Concepts for Object-Oriented Analysis and Design", *Delivrable M.a.1.2., IGLOO Project, CRIM*, January 1994

[5] H. Muller, Ronald Janmaat, "Service/Feature Interaction Resolution Framework (Guidelines)," TINA-92 Workshop

[6] Subodh Bapat, "Richer Modeling Semantics for Management Information", *Integrated Network Management, III,* eds. H.-G. Hegering and Y. Yemini, Elsevier Science Publishers B.V. (North-Holland) 1993, IFIP, pp. 15-28

[7] June Power, "The Object in Perspective", *OOPS Messenger,* vol. 4, no 4, October 1993, pp. 28-32

[8] R. Davis, R.G. Smith, "Negotiation as a Methaphor for Distributed Problem Solving," *Artificial Intelligence*, vol. 20, no. 1, 1993, pp. 63-109

[9] Petre Dini, Gregor v. Bochmann, "Design of a Configuration Management Subsystem Based on OSI/ISO Approaches, Incorporating Self-Control Mechanisms to Guarantee the Survivability of the Networks", *IGLOO Project, Deliverable A.f.2. NSERC*, CRIM, Montreal, 1994

[10] ISO/IEC JTC 1/SC 21/ DIS 7498-1, "Basic Reference Model," 1993, *Draft International Standard*

[11] ISO/]IEC JTC SC21 N8125, "ISO - Basic reference Model of Open Distributed Processing," *Draft Recommendation X.903*, June 1993

[12] Gregor v. Bochmann, Proposal of the IGLOO Project, Technical Report, CRIM, 1992

[13] *** *Proceedings of 14-th International Switching Symposium* (ISS XIV), Yokohama, Japan, 1992

[14] David Lorge Parnas, "On the Criteria to Be Used in Decomposing Systems into Modules," *Communications of the ACM*, vol. 5, no. 12, pp. 1053-1058, December 1972

[15] Petre Dini, Dunia Ramazani, Gregor v. Bochmann, 'DEF Model for Causality Design in Object-Oriented Distributed Systems", *Technical Report, IGLOO Project*, CRIM, Montréal, 1995

[16] Ronald C. van Wuijswinkel, Marc F. Witteman, "Testing Using Telecommunications Management", *IWPTS VII, 7-th IFIP WG6.1 International Workshop on Protocol Test Systems*, November 8-10, 1994, Tokyo, Japan

[17] Petre Dini, Gregor v. Bochmann, "Automatic Reconfiguration Management Using the Configuration State", *Technical Report, IGLOO Project*, CRIM, Montreal, 1995, Canada

[18] ISO/IEC JTC 1/SC 21 N 6356, *"State Management Function," final text of DIS 10164 (Recommendation X.731)*, 1991-10-15

[19] Derek Coleman, Patrick Arnold, Stephanie Bodoff, Chris Dollin, Helena Gilchrist, Fiona Hayes, Paul Jeremaes, "Object-Oriented Development: The Fusion Method," *Prentice Hall, Englewood Cliffs*, New Jersey 07632

A New Proposal for Feature Interaction Detection and Elimination

Yasuro KAWARASAKI, Tadashi OHTA
ATR Communication Systems Research Laboratories
2-2 ,Hikari-dai, Seika-cho, Soraku-gun, Kyoto 619-02, Japan

Abstract. This paper proposes a new method for detecting and eliminating feature interactions. The authors are conducting research into feature interaction detection by using production rules based on the fact that telecommunications service specifications can be expressed as finite state machines (FSMs). We have already developed new techniques for new FSM related problems found in research conducted after the last workshop, problems with terms recognized in the last workshop, explosion problems concerning resources (processor time, memory space, etc.) in interaction detection, and reduction of redundancies in interaction detection. With these new techniques, feature interaction detection and elimination at the design stage have become much more efficient.

1. Introduction

This paper proposes a new method to detect and eliminate feature interactions resulting from research undertaken after the last workshop.

Service interactions are detected in one of three stages in the software development process: the service specification design stage, the program manufacturing stage and the program execution stage. Therefore feature interactions must be eliminated in one of those stages.

It is a well-known fact that it takes from ten to a hundred times more man-hours to modify at the development stage than it does at the design stage. Worse still, if an interaction is detected during the actual provision of a service, the inconvenience caused to users might be huge. The worst case cause might serious damage to social activities. Consequently, if at all possible, it is best to detect and eliminate feature interactions at the design stage, especially when a mere adjustment in the specifications is necessary.

There are various types of interactions in which the appropriate method of eliminating each one of them will vary depending on the conditions present when the corresponding service is being provided. In such cases the interactions are each detected at the service execution stage, and eliminated. It is obvious, however, that knowledge is required as to when these types of interactions occur and how to eliminate them. To avoid this, it is immensely important to detect and eliminate interactions at the design stage.

The authors presented a classification of detection and elimination techniques at the service design stage as finite state machine (FSM) problems at the last workshop. This classification was based on the reasoning that service specifications can be described as FSMs. At that time, we divided the FSM problems into logical type and semantic type, and proposed detection and elimination methods for several of them.

In subsequent research we designed detection and elimination techniques for several more logical problems: lost states, lost transitions, and illegal transitions. We also developed techniques to reduce the redundancies of interaction detection as a whole. We applied

Petri nets to reduce the explosion of requirements for resources (processor time, memory space, etc.). On those terminological issues recognized as problematic in the last workshop, we developed a method for detecting a synonym that differs in the service specification but that is identical in interaction detection. With this new method, interaction detection and elimination at the design stage have become much more efficient.

The paper is structured as follows: In section 2, we discuss the new problems that are to be handled, and give our formalisms for these problems. In section 3, we describe techniques capable of solving the problems. In section 4, we describe an application of Petri nets that improves existing techniques. In section 5, we describe a reduction of redundancies in the elimination of transitions to illegal states.

2. New Problems in Interaction Detection

2.1. Overview of Problems

The authors have proposed methods for detecting and eliminating feature interactions based on the well known fact that communications services can be expressed as FSMs. Service specifications have been written as sets of production rules, and research has been performed by investigating FSMs produced by combining multiple services. These detection and elimination methods have involved non-deterministic transitions, and transitions to undefined states (transitions to abnormal states).

Nevertheless, there are causes of interactions that cannot be modeled using FSMs (timing, number of execution-iterations, etc.) and types of interactions based on meanings that cannot be expressed in formal specification languages. There are also other FSM-related irregularities, including illegal transitions, lost states, and lost transitions. In this section we analyze these problems, and give formal definitions for those capable of being expressed in FSMs.

2.2. Communication-Service Specification using Production Rules

Generally, when the specifications for a communications service are described in production rule notation, the condition part of a production will consist of the present service status and an event to trigger the application of the rule in question. The resultant part of the rule will be the state of the system after the rule has been applied. Production rules only express those parts of the service status being considered, so there is no guarantee that the actual status of the service (actual state) will be identical to the condition part of the production rule. A rule is applied if its condition part is included in the actual status. Furthermore, since services can be added incrementally, if a rule is created that makes a transition to a new state, then this new rule has to be given priority so that it is applied instead of the other, previous rules. For this reason, there has to be a priority application principle among the instructions when production rules are to be applied. This priority application principle should express the condition that "if there are two production rules that are logically applicable, then the rule with the more detailed state description in its condition part is to be applied". Bringing this all together for clarity, we have:
- Form of a production-rule expression:
  Current-state   Event: Next-state
- Production-rule application principle:
  Apply the production rule if:

the rule's trigger-event = the exact event that actually occurred,
and
the current status $\tilde{O}$ the actual status
• Rule priority-application principal:
In cases in which two rules are logically applicable, if
(the condition-part of Rule 1) $\tilde{O}$ (the condition-part of Rule 2),
apply Rule 2.

We also use the term "service status" here to express the combination of all individual states of the various terminals and the inter-terminal status. It is of course possible to express the transitions of the individual terminals in the production rules, but by using a variable to refer to the terminals, we can reduce the number of rules that have to be described.

2.3. Problems that cannot be expressed in FSMs

(1) Timing verification

In verifying our state transition, we have been assuming that events occur one at a time, in sequence, and that the state transition is completed at the same instant an event occurs. For this reason, we have been unable to detect interactions capable of occurring when there is a mismatch between the actual state of the network and the state a user is conscious of after he or she has caused an event to occur there. In such situations, depending on the timing of the event, it is possible for a "correct" state transition to occur that is actually at odds with the user's intentions. An example of this is the use of the "flash" command by CW and TWC members. When a member produces a "flash" because of a TWC call, but a message is received from another member immediately before that flash goes out such that the status of the network at the time the flash is processed has actually made a transition to that of a "CW call", a "CW receive" is executed contrary to the member's wishes (see Figure 1).

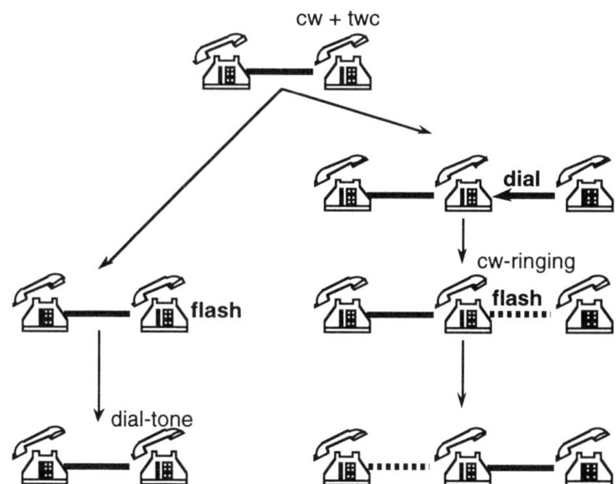

Figure 1: An example of timing -related interaction (CW and TWC)

(2) Retry verification

Specifications such as those for a retry due to a user input error, in which the system conducts a set number of repeated transitions to a given state, cannot be simply expressed as state transitions. Therefore, feature interactions resulting from differing specified numbers of retrials from service to service cannot be detected.

(3) Semantic verification of interactions

Since we had been using equivalency judgments based on event names and primitive names in our FSM-based feature interaction detection method, we were unable to detect interactions between events and primitives that were synonymous from the viewpoint of interaction detection if they were represented by different terms in the service specifications.

Semantically identical events and states are shown in Figure 2.

Figure 2: Synonyms in interaction detection

2.4. FSM problems

(1) The occurrence of illegal transitions

An "illegal transition" is a new transition resulting from combination of multiple services, between states already existing before the services were combined. We do not use the term to include new states (nor their accompanying new transitions) which can arise from the combination of services, since these new states are detectable through our previously proposed "transition to an undefined state."

The formal definition of an "illegal transition", when specifications are written in the form of production rules, is given below.

Assume that, in combining two services, A and B, there are two rules that could logically be applied in the state S_{Ai} of service A , rule r_A of service A, and rule r_B of service B. Also assume that the result of applying these rules would be transitions to states S_{Aj} and S_{Ak}, respectively. Let e_A and e_B be the events corresponding to rule r_A of service A and rule r_B of service B, respectively. If a rule identical to r_B is already included in service A, no new transition will occur. If there is no such rule included in service A, we can classify the types of things that can occur into three categories according to the results of the transition states of the various rules:

1) If there is a rule r_A that has the same event as r_B's, and if the result of transition states S_{Aj} and S_{Ak} are the same, then the rule used for the transition has changed, but the pre-transition and post-transition states are both the same as they were before the two services were combined, so no "illegal transition" occurs.

2) If there is a rule r_A that has the same event as r_B's, and if the result of transition states S_{Aj} and S_{Ak} are different, then the pre-transition and post-transition states are different from what they were before the two services were combined, so an "illegal transition" occurs.

3) If there is no rule r_A that has the same event as r_B's, the event acting as the trigger is different, so an "illegal transition" occurs.

These four types of state transitions are shown in Figure 3.

In the above discussion we limited our explanation to the combining of two services, but the same logic applies equally to the combining of an arbitrary number of services.

(2) Lost states

"Lost states" refers to states having been defined in services before being lost due to the combining of those services. When states are lost, unfortunately, the transitions to those states are also lost. Consequently it will be sufficient for our purpose to include our consideration of them in our discussions on the detection of (3) Lost transitions, below.

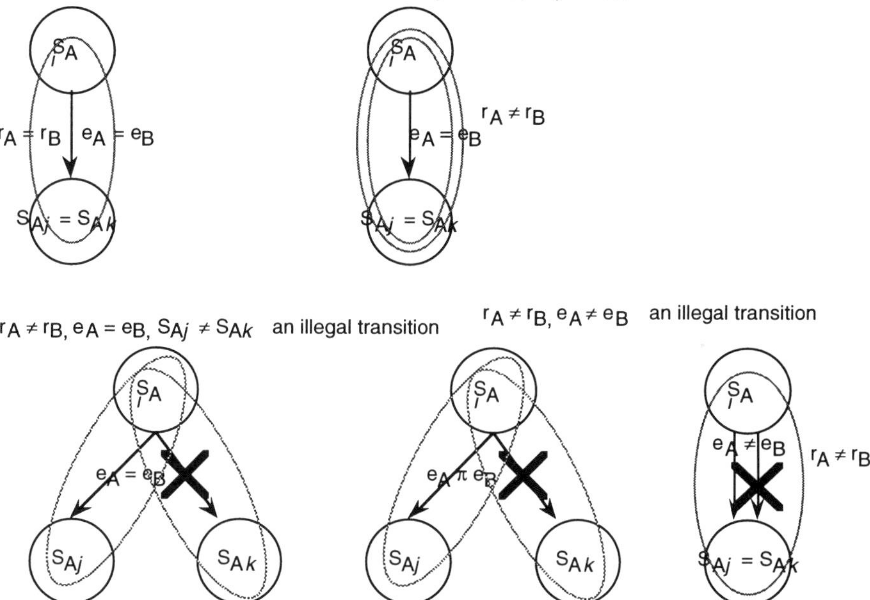

Figure 3: Four types of state transitions

(3) Lost transitions

"Lost transitions" refers to state transitions having existed in services before being lost due to the combining of those services.

The formal definition of a "lost transition", when specifications are written in the form of production rules, is given below.

When rule r_B of service B is applied with greater priority than rule r_A of service A, the transition of rule r_A is lost.

Figure 4 shows an example of an interaction between an emergency telephone service and a three-party conference-call service.

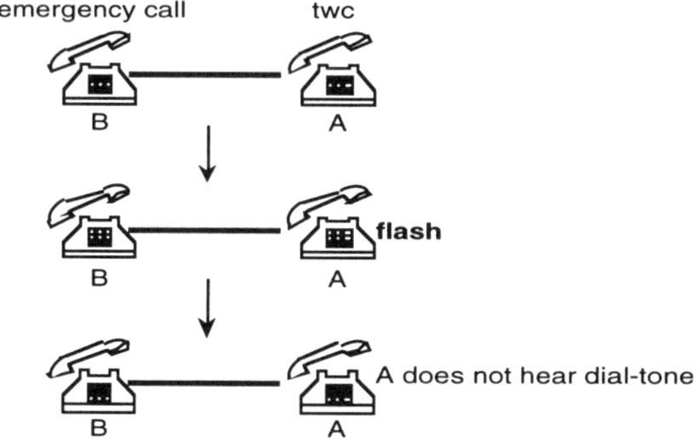

Figure 4: An example of lost states

3. A new detection method

3.1. Problems that cannot be expressed using FSMs

3.1.1. Timing verification

As we mentioned in section 2.3, a timing-related interaction occurs when a user triggers an event on the basis of his or her knowledge of the state of the system, and that knowledge does not match the actual state of the network. Consequently, if the event can produce a transition in other states from the one in which it was actually triggered, the possibility exists that an inter-service timing-related interaction will occur. The user will be aware, however, of state transitions triggered by events arising at that same terminal: it will therefore be sufficient if we consider the case of transitions arising from events triggered at other terminals.

In particular, we will consider whether or not an event e, occurring in a given state can also occur in the state previous to a transition caused by another event occurring at a different terminal. In such a situation it is possible for each of the separate services to include the above transition type among its state transitions, but we must avoid considering this here, because the individual services will each have conducted its own verifications as a separate service, prior to combining with the others.

Since one cannot determine what combinations of events can occur simultaneously merely by looking at the specification production-rules, there is a need for people who write service specifications to always double-check. The technique presented here is a way of assisting service-specification writers by producing lists of candidate combinations of states and events that might possibly lead to timing-related interactions.

The verification algorithm is as follows (see Figure 5):
1. Determine the state transition charts of each of the original pre-combination services.
2. Determine the combine-service state transition chart.
3. Perform the following steps on every transition of the combined service (where S_i is the pre-transition state, e is the event, and A is the terminal at which the event occurs):
3-1. Search for transitions resulting in S_i in the case where the terminal at which the event occurred is not A.
3-2. Determine the states sj of the transition found at step 3-1.
3-3. Check whether a transition with event e and the event that occurred at terminal A exists from S_j
3-4. If such a transition exists, check whether that transition has been included in the pre-combination state transition chart of the selected service, and if it has not, specify it as a timing-related interaction candidate.
3-5. Repeat the same process for transitions resulting in S_j. At this point, if all of the transitions resulting in state Si have been triggered by events occurring in terminals other than A, this step results in a move back to the initial state.
3-6. If there are multiple transitions satisfying the search in 3-1, repeat the same process for those remaining.

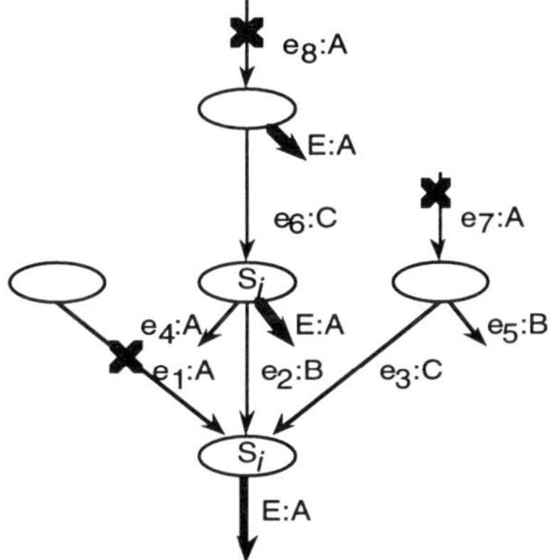

Figure 5: Timing related verification

3.1.2. Retry verification

In production rule type specifications, there is no way to directly indicate situations such as "retry", in which transitions to a given state are repeated a set number of times.

The number of retries must be defined separately from the transition rules. As a consequence, it will be possible to conduct retry verification by comparing the number of retry definitions.

3.1.3. Semantic verification of interactions

At the last workshop, we proposed a "concept model" for switching functions, as a way of handling double-definition problems in terminologies. With this method, information related to events and primitives can be similarly represented.

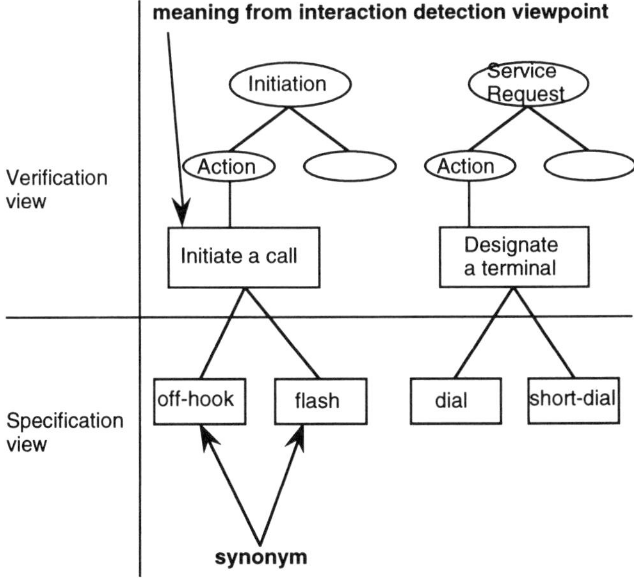

Figure 6: Concept model for semantic verification of interaction

For example, a "dial" event can be represented as the leaves of a tree structure, labeled "service selection", "selection of target for communication", and "selection of terminal to connect with". Consequently, events or primitives that have the same meanings from the point of view of verification, but which are referred to with different terms in the service specifications, are assigned to the same bottom-most concept in the tree (in the case of an event, "selection of terminal to connect with", for example) (see Figure 6).

Thus, in our previously proposed interaction-detection technique, if we use bottom-most concepts from our concept model in the place of event names, primitive names, etc., testing becomes possible for events and primitives that have the same meanings, even if they are referred to using different terms.

3.2. FSM problems

3.2.1. The occurrence of illegal transitions

In order to test for illegal transitions, transition related information is expressed in the form: pre-transition state, rule, post-transition state. Thus, a transition in service A becomes (S_{Ai}, r_A, S_{Aj}), and a transition in system B becomes (S_{Bi}, r_B, S_{Ak}). If we express the analysis in section 2.2.2(1) in this form, the case in which a rule identical to r_B is included in service A becomes:

$$r_A = r_B \qquad \text{(not an illegal transition)};$$

the case in which there is no such identical rule in service A becomes:

1) if a rule r_A exists with the same event as r_B's, and the next states S_{Aj} and S_{Ak} are identical,

$$r_A \neq r_B, e_A = e_B \text{ and } S_{Aj} = S_{Ak} \qquad \text{(not an illegal transition)},$$

2) if a rule r_A exists with the same event as r_B's, and the next states S_{Aj} and S_{Ak} are different,

$$r_A \neq r_B \text{ and } e_A = e_B \text{ and } S_{Aj} \neq S_{Ak} \qquad \text{(an illegal transition)},$$

3) if no rule r_A exists with the same event as r_B's,

$$r_A \neq r_B \text{ and } e_A \neq e_B \qquad \text{(an illegal transition)}.$$

Therefore, in order to detect illegal transitions, all we have to do is search for transitions in combined services for which $r_A \neq r_B, e_A = e_B$ and $S_{Aj} \neq S_{Ak}$ or $r_A \neq r_B$ and $e_A \neq e_B$. This can be done using the following algorithm:

1) Acquire a list of the set of transitions (pre-transition state, rule, post-transition state) of the individual services.
2) Acquire a list of the set of transitions in the combined service
3) Repeat the following process for the individual services:
   3-1. Select those transitions from the combined service whose pre-transition states are found in the pre-transition states of any of the individual services.
   3-2. From the resulting list, delete all transitions that are included in the individual service's transitions. $(r_A = r_B)$
   3-3. Further eliminate transitions that match pre-combination transitions that have the same post-transition states and the same events in their rules. $(r_A \neq r_B, e_A = e_B$ and $S_{Aj} = S_{Ak})$
   3-4. Any transitions left after the above process are illegal transitions.

3.2.2. Lost transitions

Lost transitions can be detected using the following algorithm:

1) Create sets R_A and R_B of the rules that operated in separate pre-combination services.

2) Select those members of R_A and R_B that are related by the priority principle

3) Compare the states of the transitions given priority to those of the transitions not given priority; in cases where these two states are different, a transition will be lost.

4. The state reproduction resource problem

In methods of testing for interaction using state transitions, it is necessary to reproduce every possible state that can occur in a service in order to conduct the verification. With complex services, the number of states that must be reproduced increases explosively, and, as a result, the resources needed to reproduce those states (time and memory-space) also quickly become enormous.

4.1. Petri nets as a method of reducing state production resource requirements

Here we present a technique for avoiding the explosive increase in state production resources (time and memory-space) that accompanies increases in service complexity. There is a formal specification language called the Petri net. By using Petri nets it is possible to verify reachability, reversibility, limitedness, and activeness through static analysis. The places and transitions of Petri nets correspond to the states and events of FSMs. An input arc to a transition corresponds to the present state of a rule, and an output arc corresponds to the post-rule state (see Figure 7).

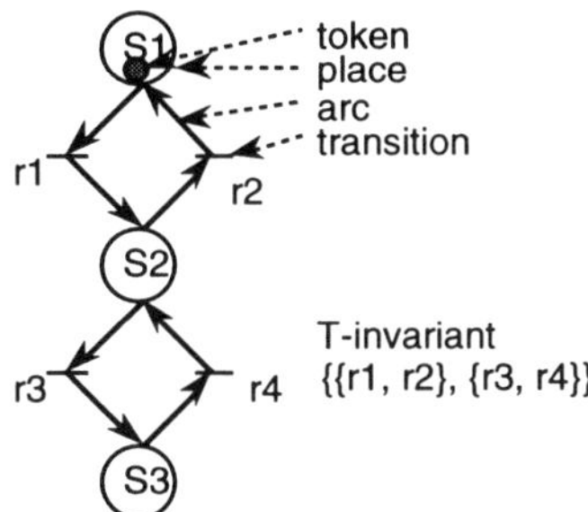

Figure 7: Petri net and T-invariant

The tokens of a Petri net can be thought of as corresponding to the terminals of a communications service . For our testing purposes, we need to differentiate between terminals, so we use a net with colored tokens, and colored arcs. We call this sort of Petri net a "colored Petri net". In a colored Petri net, transitions only fire when there are colored tokens corresponding to the colors of the individual arcs. Petri-net transitions can be thought of as analogous to FSM transitions.

The reachability, reversibility, limitedness, and activeness characteristics of Petri nets correspond to the reachability and reversibility of FSM states, the finiteness of the number of states, and the quality of non-producing deadlocks, respectively. Thus, if we can convert production rules into a Petri net, we can verify the above characteristics by analyzing that net.

We will now discuss the reversibility characteristic mentioned above. If we apply the FSM reversibility concept to communications service specifications, it means that it is always possible to reach the idle state from any arbitrary state that can be reached from the idle state. The event occurrence chain in this situation can be expressed queue style as a Petri-net transition triggering series, known as a T-invariant. Consequently, in verifying the reversibility of an FSM, it is sufficient if we reproduce states according to the event-occurrence chain represented as a T-invariant. We can thus avoid the reproduction of states due to the application of unnecessary production rules.

The procedure for reproducing states using a Petri net is described below.

(1) Converting production-rule type specifications into a Petri net

Conversion from production rules into a colored Petri net can be conducted according to the following procedure: (see Figure 8)

1. Convert events into transitions:

   Convert the event of each rule into one transition.

2. Convert states into places:

   Convert the states included in each rule into places. In this step discrimination between variables is handled by coloring the arcs, so only one copy of each state is produced.

3. Color the arcs:

   Assign one color to each of the terminals included in each rule. Link places and transitions using arcs of the color assigned to the variable included in the state corresponding to each place.

4. Gather together the colored Petri nets, which were produced in steps 1-3, that correspond to one rule:

   Since colored Petri nets corresponding to each rule were produced in steps 1-3, we can create a single colored Petri net corresponding to the communications service specifications by simply taking identical places that exist in different nets and replacing them with a single common place.

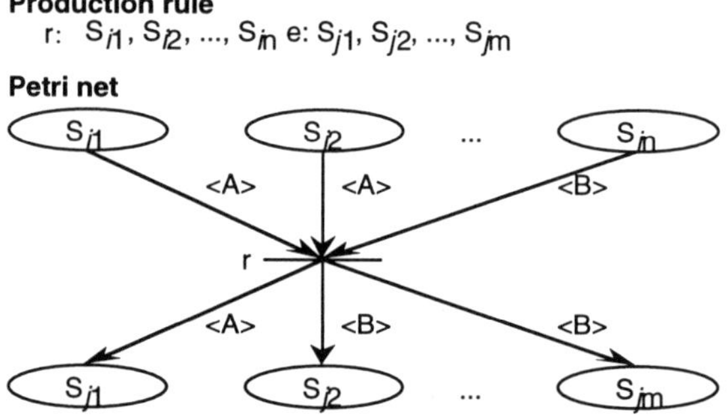

Figure 8: Conversion from production rules into a colored Petri net

(2) Converting colored Petri nets into colorless Petri nets

Since no efficient method of determining the T-invariant of a colored Petri net has yet been discovered, we will first convert to a colorless Petri net, and then seek the T-invariant in that colorless net. It is possible to convert a colored Petri net to a colorless Petri net by simply eliminating the color-assignment information corresponding to each arc; however, the conditions of a colorless Petri net are more lenient than those of a colored Petri net; cases exist in which the T-invariant of a colorless Petri net is not the T-invariant of the colored Petri net.

(3) Calculation of the T-invariant of a colorless Petri net

The T-invariant can be determined by matrix calculations[8].

(4) Producing a state transition chain on the basis of a T-invariant

Since, as mentioned in section (2) above, there are cases in which the T-invariant of a colorless Petri net is not the T-invariant of the colored Petri net, some of the T-invariants sought will be redundant with respect to the production-rule type specifications. Nonetheless, compared to conducting tests by reproducing every state within the scope of possible state transitions, the scope is much more limited by the colorless Petri-net T-invariants, so this method is extremely effective.

In addition, judgment as to whether a given T-invariant is redundant or not with respect to the production-rule type specifications can be made on the basis of whether or not the transition corresponding to the calculated T-invariant is possible according to the production rules. When doing this, it is possible to determine the order in which the rules are applied at the same time — this is something that cannot be determined using T-invariants.

(5) State production based on a calculated state transition chain

We seek the upper limit on the number of times each rule is applied using the multiple rule-application chains determined in section (4). As in the traditional state production method, we apply the rules in order, starting at the initial state. On two points, however, our method differs from the traditional one: we apply only those rules that agree with the order of the rule-application chain, and we do not apply any rule more than its specified number of applications. This allows us to eliminate unnecessary rule-applications.

4.2. Evaluation

While the number of reachable states is not reduced using the present technique, it does allow a reduction in the number of rule-applications needed to reach those states. We tested the method by producing every possible state that can occur using two terminals with the POTS service.

When the T-invariant was not used, the number of rule-application trials was 72, whereas when the T-invariant was used, the number was 34 — a reduction by more than half. This improvement represents a marked increase in the use ability of FSM-based feature interaction detection.

5 Reduction of redundancies in interaction detection

5.1. The redundancy of the specified interaction-elimination proposal

We have proposed two techniques for assisting in the elimination of feature interactions so far: a technique that assists in eliminating indeterminacy, and a technique that assists in eliminating transitions to undefined states. In these techniques, some conditions were loosened to produce more candidates for elimination, so that no candidates would be missed. For this reason, there are state transition rules specified that have no meaning for a communications service.

In the latter technique, the present state is set to the state effective at the time a transition to an undefined state is detected, the event is set to the detected event, and the subsequent state is set by selecting the states of each service that conform to the following regulations, thus producing candidates for elimination:

1. The states that result from applying the rules produced by combining the two can be expressed as the union of the sets of states in each of the individual pre-combination services.
2. Transitions related to the terminal for which an event occurs are executable terminal-state transitions, when the rules of the individual services are applied separately.
3. Transitions related to terminals for which no event occurs are executable transitions when an arbitrary event occurs at one of these terminals, when the rules of the individual services are applied separately.

When applying condition 3, candidates for elimination are created on the basis of the occurrence of an arbitrary event, so some transitions to states for which there is no transition when an event e occurs in an individual service may be included among them. For that very reason — that these candidates include transitions to states for which there is no transition, in an individual service — they are often meaningless in the context of a communications service; however, there are also cases in which they are valid candidates, because of the grouping of the services in the combined-service structure. Consequently, rather than simply eliminating those candidates that include transitions to transitionless states in an individual service, what is needed is a system that will assist the service-specification coder by giving such candidates a lower priority of candidacy, or use other such methods of tightening-up the candidate list.

5.2. Countermeasure

The problem is this: candidates are listed without separating them according to whether or not they include transitions to states for which there is no transition, in an individual service, at the time of the occurrence of event e. Therefore, by comparing the transitions that occur when event e occurs in each individual service for each primitive in the subsequent-state of a candidate for elimination, and then by classifying these into candidates that do not include transitions to states in individual services for which there are no transitions, and those that do; and by indicating the higher priority of the former, the problem is solved.

6. Conclusion

We have presented techniques for detecting the production of illegal transitions, lost states, and lost transitions, for use in interaction detection with specifications written in production rules. In addition, we have presented techniques for dealing with interactions that do not come to light in FSMs: timing related interactions, interactions related to the number of retries, and interactions resulting from the use of different words in service specifications, that are semantically identical from the point of view of interaction detection.

Tasks that await our attention are a method for the automatic conversion, into our concept model, of new words that have become necessary for a given meaning; and further reductions in the number of states that must be reproduced for detection. Furthermore, for conformity with existing development methods, there is the task of creating an automatic method of converting from the service specification languages used by existing services into STR. For the latter, definite prospects do exist for automatic conversion in the cases of MSCs (Message Sequence Charts) and IN service-customization languages.

Acknowledgments

The authors would like to thank Dr. K. Habara and Dr. N. Terashima, chairman and president of ATR Communication Systems Laboratories, respectively, for their encouragement and support. The authors also thank the members of the Communication Software Department for their helpful comments and discussions.

References

[1] T. F. Browen, et al., "The Feature Interaction Problem in Telecommunications Systems," Proc. of 7th IEE Int. Conf. on Soft. Eng. for Telecom. Systems, pp.59-62, July, 1989

[2] E. J. Cameron, et al., "A Feature Interaction Benchmark for IN and Beyond," Proc. of Workshop on Feture Interactions in Telecommunications Systems, pp.1-23, 1994

[3] A. Gammelgaard, at el., "Interaction detection a logical approach," Proc. of Workshop on Feature Interactions in Telecommunications Systems, pp.178-196, 1994

[4] Y. Harada, at el., "A Conflict Detection Method for Telecommunication Billing Specification Descriptions," IJCA Tran., 34, 5, pp.1064-1073, 1993

[5] Y. Hirakawa, at el., "Behavior Description for a System which Consists of an Infinite Number of Processes," BILKENT Int. Conf. on New Trends in Communication, Control, and Signal Proc., pp.59-68, 1990

[6] Y. Inoue, at el., "Method for Supporting Detection and Elimination of Feature Interaction in a Telecommunication System," Proc. of Workshop on Feture Interactions in Telecommunications Systems, pp.61-81, 1992

[7] E. Kuisch, et al., "A Practical Approach Towards Service Interactions," Proc. of Int. Workshop on Feature Interactions in Telecommunications Software Systems, pp.41-59, July, 1992

[8] P. M. Merlin, "A Methodology for the Design and Implimentation of Communication Protocols," IEEE Trans. Commun., 24, 6, pp. 614-621, 1976

[9] T. Ohta, at el., "Acquisition of Service Specifications in Two Stages and Detection/Resolution of Feature Interactions," Proc. of TINA'93, 2, pp.173-187, 1993

[10] T. Ohta, at el., "Classification, Detection and Resolution of Service Interaction in Telecommunication Services," Proc. of Workshop on Feture Interactions in Telecommunications Systems, pp.60-72, 1994

[11] W. Reisig, "Petri Nets," Berlin Heidelberg: Springer-Verlag, 1995

[12] Y. Wakahara, et al., "A Model for Detecting Interactions Among Service and Service Features," Proc. of Int. Workshop on Feature Interactions in Telecommunications Software Systems, pp.82-94, July, 1992

Representing and Verifying Intentions in Telephony Features Using Abstract Data Types

Bernard Stepien and Luigi Logrippo
Telecommunications Software Engineering Research Group
Department of Computer Science, University of Ottawa
Ottawa, Ont. Canada, K1M 6N5
(bernard | luigi)@csi.uottawa.ca

Abstract. Feature intentions describe the intended behavior of telephony features. A method for formally specifying feature intentions by abstract data types is given. Further, a method for detecting violation of certain types of feature intentions at the design stage is provided. Specification languages considered are SDL, Prolog, and LOTOS. If the specification is in LOTOS, detection can be helped by the use of goal-oriented execution. Examples of specification and detection are provided, namely involving Originating Call Screening and Call Forwarding.

1. Introduction

[CGL94] provide a survey of the many types of feature interactions possible in telecommunications systems, and propose some categorization schemes. In one of these they distinguish among three main causes of feature interactions: violation of feature assumptions, limitations on network support, and intrinsic problems in distributed systems. [BW94] categorize the different techniques to deal with the problem in the three following categories: avoidance, detection, and resolution. In this paper, we propose a new detection technique, which can be used at the design and specification stage, to find interactions caused by violations of feature assumptions and by intrinsic problems in distributed systems.

Feature interactions are not necessarily a problem. Some are intended and some are undesirable. In this context, it is difficult sometimes to base detection on service specifications alone, that cover only the procedural or technological aspects of features. In many instances, features are meant to implement some subscriber's wishes or intentions. An undesired feature interaction can thus be defined as a violation of intentions. This concept is reflected by many authors in their work at various levels of their analysis. For example [LL94] distinguish between procedural and behavioral specifications of features and [GV94] distinguish between technological and policy features. In the first case, the authors use procedural and behavioral specifications concurrently to detect feature interactions. In the second case, the authors separate feature interactions into two groups and decide to address the policy features that reveal mainly intention violations. Technological feature interactions are detectable by inspection of the procedural specification. Feature interactions resulting from violations of intentions however cannot be always detected from such inspection alone because they depend on the intentions of the user (behaviors or policies). Consequently, we need to be able to describe explicitly and separately such intentions and verify them against the specification of the various features present in a system. Some feature interactions resulting from scattering in a distributed system can also be detected using this method. The two papers mentioned so far have clearly separated the actual specification of intentions from the feature interaction detection mechanism. While the basic principles are similar, their implementations were very different. [LL94] use temporal logic while [GV94] use proposals or goals in a Prolog form to represent user intentions.

Intentions have also been described as *invariant* properties [Zav93], which must hold throughout the system's lifetime.

In this work, we propose a technique for feature intention violation detection at the design stage, based on Formal Description Techniques (FDT) that use Abstract Data Types (ADT) to represent intentions. We show how the technique can be used to detect feature interactions in association with various standard procedural formal description techniques. LOTOS appears to be the most suitable one, but SDL is also suitable to a certain extent. A more generic technique based on Prolog programming is also considered. We stress some benefits of the systematic specification and detection approach that is made possible by FDTs.

1.1 Definition of a generic intention representation

We start by constructing an executable formal model (or prototype) of the system with features. This formal model can be built in any of the languages mentioned above. Intentions (i.e. invariants) are represented in such a way that when a feature is activated, they can be verified for every subsequent action. When verification fails, the invariant has become false, so a flag is raised. Consequently, we need a generic mechanism to verify simultaneously many different features that might have been activated during the execution of a specification that represents a connection attempt between two subscribers within their own respective environments. The principle of a generic data type representation is useful, since the specification of the features and of their interaction detection mechanism is isolated in the data part of the specification. Our technique is related to the one of [GV94]. Although their paper deals with run-time detection, while we deal with detection at the design stage, we do detection by executing the specification. Furthermore, we use a flat, or tabular, representation, as opposed to the hierarchical system of [GV94]. The latter was motivated by a resolution technique, which is not among our goals.

2. Representing intentions using Abstract Data Types

In order to define a generic representation and verification mechanism to support the specification of a variety of features, we need first to determine the categories of data and the rules associated with features. Telephone systems operate with a limited set of data or rules: phone numbers, signals, feature names, and databases of various types of names and phone numbers. Rules usually are restriction sets such as in the *Originate Call Screening* feature. In its ADT representation, a feature is an operation that has the above mentioned data as its domain, and uses either a substitution or a restriction rule to determine its range.

The intention of a feature can be abstracted to rules with specific combinations of the four basic classes of data delimited above. So we can represent an intention as an operation of boolean result that indicates whether a given combination of the basic data involved in an operation is allowed.

intention: Fid, partyRole, operation, Restriction_set -> Bool

For example the originate call screening (OCS) prohibits any connection with a number that is in the screening list. This can be formulated thus:

The number that has the called role shall not be connected if it is in the screening list.

The above example can thus be expressed with the following equation:

intention(F_{ocs}, called(N), connect,L) = N NotIn L;

where F_{ocs} identifies the feature Originating Call Screening, N the phone number involved in a called role in operation *connect* and L is the restriction set that in this case is a screening list, *N NotIn L* is a boolean expression that verifies if the number N is not found in the restriction set L, which would mean that the intention is verified for this particular operation. Note this set will be used in different ways.

It is clear from this definition that equations have to be provided for each type of operation, including the operations that are specific only to certain features. Similarly, the interaction detection methods used in [BA94] uses a tabular notation, where analysis of these tables against actual features can lead to interaction detection. Our equations play the same role as the tables used in their method. However, the representation of feature intentions is only a first step toward feature interaction detection. In our method, only the actual simulation of two simultaneous features can verify if an interaction occurs. However with intentions clearly defined, the detection of an interaction can be automated (at least in part) and is no longer subject to the designer's judgement. Also, it is now clear that the specification of an intention consists in specifying the constraints on a set of operations. For the application of our method, two guidelines must be respected:

• Intentions of a feature are described independently of other features. This means that the specification of an intention on a specific operation is done solely in consideration of the given feature. There is no consideration of potential interactions with another feature at this stage. This is important especially in the context of a multi-vendor environment where characteristics of a feature of a vendor may not be known to another vendor. Also, from a design point of view, this characteristic allows the feature provider to add new features without having to understand their behavior in combination with other features, as already mentioned in [GV94].

• Intentions of a feature must be described for every operation that exists in the system regardless of which feature is actually used. This means that operations must be considered as belonging to some sort of generic pool of operations that can be used by various features. This, even if some operations were created specifically for a feature such as for example Originating Number Display. While originally an operation may have been created for a feature, chances are that it will be re-used by other features and thus become more generic. This convention is a result of the formalism we are using, because with ADTs the equation set must be complete and convergent.

For implementation purposes we have chosen to represent intentions as ADT operations that actually detect the intention's violation. We thus define operation *violates* as having the same domain as the intention operation defined above. The range of this operator is also a boolean value that indicates whether the intention has been violated or not. However, the boolean expression is reversed. For example, here we are saying that, in the case of F_{ocs}, a violation occurs if the number involved in operation *connect* is found in the restriction list:

$$\textbf{violates(Focs, called(N), connect,L) = N IsIn L;}$$

2.1 Examples

The use of our representation of intentions can be illustrated in three examples that cover the two main categories of violation of assumptions and problems in distributed systems defined in [CGL94]. The remaining category of limitations in network support is not addressed here because it is mainly the result of problems linked to ambiguities that have been covered with the backward reasoning methods in [SL94].

2.1.1 Originate Call Screening

This feature is known to interact with the call forward feature and this is an example of interaction caused by intrinsic problems in distributed systems. Here is the full example of the specification of the intention violation for the feature, where L is the screening list:

```
type intentions is Boolean, features, signals, roles, number_set
      sorts intention
      opns
            violates: feature, role, signals, number_set -> Bool
      eqns
            forall N:number, S: signals, L:number_set
            ofsort Bool
                  violates(Focs,called(N),connect,L) = N IsIn L;
                  violates(Focs,called(N),off_hook,L) = false;
                  violates(Focs,called(N),tone,L) = false;
                  violates(Focs,called(N),dial,L) = false;
                  violates(Focs,called(N),conreq,L) = false;
                  violates(Focs,called(N),detect_forward,L) = false;
                  violates(Focs,called(N),connect_refused,L) = false;
                  violates(Focs,called(N),ring,L) = false;
                  violates(Focs,called(N),answer,L) = N IsIn L;
                  violates(Focs,called(N),talk,L) = N IsIn L;
                  violates(Focs,caller(N),S,L) = false;
endtype
```

We have here some rules for which the outcome is immediately determined (false) while for others we need to further evaluate if a number belongs to the restriction set. We can also observe that the F_{ocs} rules are complex only for the called role because the restrictions apply for the called party, while there is only one catch-all rule for the caller role that always evaluates to *false*.

2.1.2 Call Forward

Another interesting example related to intrinsic problems in distributed systems is loop or livelock detection for the call forward feature. These loops are the result of personalized instantiation by the subscriber of the feature. Here the restriction set L contains the phone number that has activated its call forward feature, to prevent circular call forwarding. The set of equations is similar to the one we have seen for originate call screening:

eqns
       forall N:number, S: signals, L:number_set
       ofsort Bool
             violates(Fcfw,called(N),connect,L) = N IsIn L;
             violates(Fcfw,called(N),off_hook,L) = false;
             violates(Fcfw,called(N),tone,L) = false;
             violates(Fcfw,called(N),dial,L) = false;
             violates(Fcfw,called(N),conreq,L) = false;
             violates(Fcfw,called(N),detect_forward,L) = false;
             violates(Fcfw,called(N),connect_refused,L) = false;
             violates(Fcfw,called(N),ring,L) = N IsIn L;
             violates(Fcfw,called(N),answer,L) = N IsIn L;
             violates(Fcfw,called(N),talk,L) = N IsIn L;
             violates(Fcfw,caller(N),S,L) = false;

2.1.3 Unlisted number

Finally, the detection of the interaction between the unlisted number feature and the calling number display feature that is an example of violation of assumptions is feasible when specifying the intention violation rules for the unlisted number only. In this case, the only operation that would cause a violation is a display operation. Consequently there are only two equations required to describe these intentions:

eqns
       forall OP:signal, ROLE:role, L:number_set
       ofsort Bool

             violates(Fuln, calling(N), display, L) = N IsIN L;
             OP ne display => violates(Fuln, ROLE, OP, L) = false;

Here the restriction list L contains the number that has the unlisted number feature activated. If this number is involved in a display operation in a calling role, this means that there is a violation of intentions.

As a final remark on these examples, we note that one must be careful about how rules that detect violations of intentions are defined, because there could be interactions between the rules themselves, both inside a single feature, and among features. For example, one may decide that ringing a number that is on the originate call screening feature restriction list is a violation. This happens to be true when the OCS feature is considered in isolation from other features. But if that same called number has a call forward feature that is activated, then the caller should not be restricted to hop to another number that is not on the screening list. This however introduces a delicate intention definition problem, well-known in this research area: should we consider numbers or subscribers as a decision criterion. In the second case, if the intention is really to prevent talking to a given subscriber, then ringing and call forwarding should also be disallowed in the originate call screening rules.

2.1.4 Other languages

So far, the ADTs were shown in LOTOS syntax. However SDL syntax is very similar, only keywords and delimiters change, and the semantics is identical.

Here is the SDL specifications of intentions for F_{ocs}:

newtype *intentions*
      operators
            violates: feature, role, signals, number_set -> Bool;
      axioms
            for all N **in** number, S **in** signals, L **in** number_set
            ofsort Bool
                  violates(Focs,called(N),connect,L) == N IsIn L;
                  violates(Focs,called(N),off_hook,L) == false;
                  violates(Focs,called(N),tone,L) == false;
                  violates(Focs,called(N),dial,L) == false;
                  violates(Focs,called(N),conreq,L) == false;
                  violates(Focs,called(N),detect_forward,L) == false;
                  violates(Focs,called(N),connect_refused,L) == false;
                  violates(Focs,called(N),ring,L) == false;
                  violates(Focs,called(N),answer,L) == N IsIn L;
                  violates(Focs,called(N),talk,L) == N IsIn L;
                  violates(Focs,caller(N),S,L) == false;
endnewtype intentions

Intentions can be easily represented with PROLOG clauses. The following example illustrates how we can specify the violation condition for the connect operation in a call forward feature.

      violates(cfw,called(N),connect,L):-
                              isin(N,L).
where clause isin(_,_) is defined as:

      isin(X,[X | _]):- !.
      isin(X,[_ | T]):-
                    isin(X,T).

3. Detecting violations

3.1 Basic mechanism

So far, we have seen how to specify violations. Mechanisms for detection must now be considered. Our method works by executing a formal specification (which could be called also formal prototype or formal model) of the system with features. The ADTs, included in the specification for this purpose, will check execution sequences to see whether a violation is incurred.

As mentioned earlier, violations of intentions can result from the lack of continuity that is inherent to distributed systems. A network element may forward processing to another network element without passing feature data, or the logic of a feature may be activated too early or too late or not at all when the critical data it needs becomes available. Our technique records the initial intentions of a feature and activates a mechanism that is independent of the distributed system to verify that the original intentions still hold regardless of which component of the system handles a call. The ADTs represent a kind of a central call model monitor that captures every event associated with a call, regardless of the component in which it occurs, and verifies if that event conflicts with the predefined intentions. Such a central mechanism would be impossible to implement in an actual system, but is possible in a specification.

3.2 Implementation

There are two main implementation concerns:

- minimal disruption of feature specifications.
- independence of specification style.

The first concern means that we do not want to modify the specification of the system in detail for the verification of each feature, as mentioned in [DN94]. The second concern means that we do not want to constrain the usefulness of our method to a specific structure, or style, or even specification language.

In LOTOS, when several processes are combined together by means of the parallel composition operator | |, in order for an action to execute, all processes must participate simultaneously (synchronize) in the action. Further, each process can provide its own conditions for the action to execute, and all such conditions must be true simultaneously in order for this to happen. This effect is inspired by CSP [Hoa85] and is further explained in [FLS91]. Thus the monitoring effect we want to obtain can be achieved gracefully by using an independent monitoring process in parallel with the system specification. This process will synchronize with every action occurring in the telephone system, and will check for every action whether a violation is occurring.

$$\text{telephone\_system[u,n] | | verify\_intentions[v,u,n]}$$

In SDL, this type of immediate and continuous monitoring is harder to achieve. One would need to set up some message duplication through a channel connecting a verification process.

In Prolog, one needs to add a verification clause to each execution rule.

```
exec_action(sequence(action(A,F,RL),RS),A,RS):-
            verify_intentions(A,F,RL).
```

Since LOTOS is the language in which this method appears to be most easily specified, henceforth we limit ourselves to LOTOS and we leave the other languages for further research.

We have studied two different approaches for the procedural part of the verification process. They are based on different principles. In the first one, most of the work is done by appropriate processes, while in the second one, most of the work is done by appropriate ADTs. Thus we call these methods respectively process-oriented and ADT-oriented. In both cases, we need to record the fact that a feature has been activated and we need also to verify for each subsequent action occurring in the system if an intention has been violated.

3.3 Process oriented feature intentions verification

In this method, the *verify_intentions* process spawns an independent monitoring process every time a new call is created. *verify_intentions* itself is in interleave (operator | | |) with the intention monitor, i.e. it runs in parallel with it but without participating in (synchronizing with) its actions. Each intention verification process is independent of the others. Consequently the *verify_intentions* process uses the call initiation action *off_hook* as a trigger to create instances of process *intentions_monitor.*

process *verify_intentions*[f,u,n]:noexit:=

```
    u ? N:number ! off_hook ? L:location ? ConId: ConIdent
    ;
        (
            intentions_monitor[f,u,n](ConId)
          | | |
            verify_intentions[f,u,n]
        )
endproc
```

However, a call may or may not activate a feature. Thus the process *intentions_monitor* has a normal mode of operations in the case no feature is activated, and this mode can be disabled ([> operator) at any time by a mode where a feature is activated and begins to be monitored.

process *intentions_monitor*[f,u,n](ConId:ConIdent):noexit:=

```
    normal_operation[u,n](ConId)
    [>
        activate_intention[f,u,n](ConId)

endproc
```

In order to detect the activation of a feature we need an explicit action in the telephone system specification. This action carries information such as the nature of the feature and its restriction set.
  The distinction between many instances of intention monitors corresponding to different connections in the system is achieved using a connection identifier *ConId* that is captured on the *off_hook* action and is propagated throughout the various actions belonging to the same connection.

process *activate_intention*[f,u,n](ConId:ConIdent):noexit:=

```
    f ! ConId ? Feature: feature ? SL: number_set
    ; monitor_one_feature[u,n](Feature,SL,ConId)
```

```
        | | |
        activate_intention[f,u,n](ConId)
endproc
```

The process *activate_intention* will spawn a process *monitor_one_feature* that is dedicated to monitoring one activated feature for a given connection. It also uses a recursive interleave construct to allow more than one activated feature to be monitored at the same time. Some connections will activate many different features as they progress in time. For example, using a resource oriented style [VSV91], the originate call screening feature is activated by a call initiator behavior using gate *f* to separate it well from the regular specification of the feature:

```
process phone[u,n,f] (N:number,FWD:feature_data,OCS:feature_data):noexit:=

        u ! N ! off_hook ! origin ? ConId:conIdent
        ; f ! ConId ! extract_feature(OCS) ! extract_ocs(OCS)
        ; n ! N ! tone ! origin ! ConId
        ; u ! N ! dial ? C:number ! origin ! ConId
        ; ...
endproc
```

The process *monitor_one_intention* is where the verification really occurs. It will make use of the generic *violates* operator.

```
process monitor_one_feature[u,n] (F:feature,SL:number_set,ConId:ConIdent):noexit:=
        hide v in
        n ? N:number ? S:operations ? C:number ? L:location ! ConId
        ; (
                (* a violation has been detected, display a message and stop *)
                [ violates(F,determine_role(L,N),S,SL) ] -> v ! violation ! F ! S ! ConId ; stop
                []
                (* no violation detected, move on to the next action verification *)
                [ not(violates(F,determine_role(L,N),S,SL)) ] ->
                                monitor_one_feature[u,n](F,SL,ConId)
        )
        []
        (* etc...*)
endproc
```

The first action *n ? N:number ? S:operations ? C:number ? L:location ! ConId* captures the data associated with an operation. These are *S*, the name of the operation, and *L* the location indicator (originating or terminating). Then, this data capture operation action is followed by a non deterministic choice expression that offers either an action where the evaluation of the *violates* operation is true (a violation has occurred), or a silent move represented by a recursive invocation of the *monitor_one_intention* process in the case where the evaluation of the *violates* operation is false (no violation).

Normally, in a LOTOS specification one has several action types, where an action type is determined by the gate and by the data elements offered. Since actions can synchronize only if they have the same types, for each type of action we need a verification choice construct. The first guarded expression in the construct handles a violation, while the second guarded expression represents the transition in case there was no violation.

3.4 Abstract Data Type oriented intentions verification

In the second approach, there is only one feature monitor that handles all activated features at once. Feature activations are kept in a set of activated features and every time an action is executed, the verification is performed
recursively for each activated feature contained in that set. If a violation is
detected, a message indicating the circumstances is displayed and the engine
keeps looking for more violations.

The intentions monitor consist of a non-deterministic choice between a feature activation detection action that will result in the update of the activated
features set, and an action verification process that will verify intentions for
every action occurring except for a feature activation action.

process *verify_intentions*[f,u,n](AFS: ActFeatSet):noexit:=

```
    (* feature activation detection mechanism *)
     f ? N:number ? F: feature ? SL: number_set
     ; verify_intentions[f,u,n](Insert(activatedFeature(F,N,SL),AFS))
    []
    (* action verification mechanism *)
     verify_action[f,u,n](AFS)
endproc
```

The action intentions verification process will start the recursive intentions verification for each kind of action.

process *verify_action*[f,u,n](AFS:ActFeatSet):noexit:=

```
    u ? N:number ! off_hook ? L:location ? ConId:ConIdent
    ; verify_one_intention[f,u,n](N,ConId,off_hook,L,AFS,AFS)
    []
    ... (* one similar construct for each type of action *)
endproc
```

The feature intentions verification process *verify_action* is a recursive process
that passes the data associated with an action to process *verify_one_intention*.
This process, that is in parallel with the telephone system specification, evaluates this data for each activated feature until all activated features have been
verified. It then moves on to the next action by recursively invoking process *ver-
ify_one_intention*. Thus, process *verify_one_intention* acts as a kind of list processor.

This alternate method was found to be more implementation-oriented, but
also more difficult to use with our simulation strategy (next section), thus it was
not investigated further.

4. Simulation strategies: use of goal oriented execution

So far we have presented a technique to represent feature intentions in a specification in such a way that any intention violations encountered when the feature is executed are flagged. We now discuss strategies that will lead to
detection of potential violations. To do this, one needs to execute the specification in some way.

As we have seen, the definition of intentions consists in defining equations for each operation. Looking for interactions means looking for the execution of actions which cause the execution of these operations.

Combinatorial explosion of the state space is a well-known problem that is encountered when trying to explore all paths in the specification of a complex system. Some relief strategies are known, in particular two methods were implemented in our tools: goal-oriented and backward execution. These methods help derive the execution paths necessary to reach a goal which is one of the actions or operations we are looking for. By inspecting these paths we can see which features were activated at the time the violation was detected, and conclude that an interaction between them was detected. Goal oriented execution has been introduced in [HLS93] [BE93] and backward execution, which is a variant of goal oriented execution, has been presented in [SL95] for the LOTOS paradigm. As far as we know, no such method currently exists for SDL, but the ideas presented in [DB78][Hol85] apply to finite state machines thus could be adapted to SDL.

Both methods are useful for finding execution paths leading to a specified action, which we call the *goal*. In goal-oriented execution, the execution tree is narrowed by using syntactic information that eliminates paths exploring parts of the specification where obviously the goal action cannot be found. In backward execution, the specification is executed backwards from the goal action. Both techniques are complicated by the need of taking into consideration concurrent paths.

As an example, to find a violation in the case of the Originate Call Screening feature we need to perform a goal oriented execution looking for the action *connect* for a number that is in a called role, and which is in the screening list. This can be found in the specification of a *callResponder_role* process of a phone. Going backward we find that this operation is possible only if a phone has been rung and the high-level specification shows that this can occur only if a switch has synchronized with a phone on the *ring* operation. We now inspect the specification of a switch to find out that there are two different ways a switch can ring a phone: either directly as a result of a connection request or indirectly as a result of a call forward. Going backward on these two paths, we find that a *callInitiator* phone would have had to place a connection request, that such a phone could have had an originate call screening feature activated, and that the phone that has forwarded its calls to the prohibited number could have been the one that was rung by this phone.

In this case, our goal was the action *connect* that belongs to the telephone system. Another way to perform the goal oriented execution would have been by using the violation detection action of the intentions monitoring process that is in parallel with the telephone system specification. This is a powerful and unmistakable method for violation detection, but experience has shown that it is very difficult to implement with the tools we currently have, because of combinatorial explosion. We need more advanced relief techniques to make this method workable.

The feature interaction between F_{ocs} and F_{cfw} has been detected using the University of Ottawa Goal Oriented Execution tool, which is a part of the ELUDO toolkit. The system configuration used was as follows:

```
(
    (
        phone[u,n,f](num1,fwd(none),ocs(Insert(num3,{}))) )
```

```
            |||
          phone[u,n,f](num2,fwd(num3),ocs({}))
            |||
          phone[u,n,f](num3,fwd(none),ocs({}))
        )

     |[n]|

        network[n]
     )

     ||

   verify_intentions[f,u,n]
```

In this configuration phone *num1* has F_{ocs} while phone *num2* has F_{cfw} to *num3*.

Our end goal is *v ! violation ! 'OCS' ! connect ! conn1 ! num3*. However by using only this goal, the tool runs into combinatorial explosion in trying to complete all paths that could possibly lead to this action. The search can be reduced by giving some further directions to the tool. These directions are directly related to the goal. First it is clear that we are placing a call from a phone that has its F_{ocs} activated. In our case this happens to be phone num1. We thus specify the three necessary actions for phone num1 to place a call regardless of the terminating number. This is indicated with the three actions:

```
u !num1 !off_hook !origin !conn1,
u ! num1 ! dial ? C ! origin ! conn1,
n ! num1 ! conreq ? C ! origin ? ocs ! conn1,
```

Also we know that at least two features have to be activated, first the F_{ocs} feature and then any feature. This is indicated with the two generic actions:

```
f~,
...
f~
```

Finally we need to indicate the offending action for which we know that a violation could be detected on the terminating side:

```
n ! num3 ! connect ! dest ! conn1
```

The goal definition, shown in the form of the actual command to the tool, is as follows:

Goal we want to reach:

```
[u !num1 !off_hook !origin !conn1,
 f ~,
 u ! num1 ! dial ? C ! origin ! conn1,
 n ! num1 ! conreq ? C ! origin ? ocs ! conn1,
 f ~,
 n ! num3 ! connect ! dest ! conn1,
 v ! violation ! 'OCS' ! ring ! conn1 ! num3] \\ [f, u, v].
```

where \\ [f, u, v] means that no intermediate actions involving these gates should exist in the trace (intermediate actions on gate n only are allowed).
　　The resulting trace is:

--> u !num1:number !off_hook:operations !origin:location !conn1:ConIdent line(s): [303,214]
--> f !conn1:ConIdent !OCS:feature !Insert(num3, {}):number_set line(s): [345,215]
--> n !num1:number !tone:operations !origin:location !conn1:ConIdent line(s): [377,262,216]
--> u !num1:number !dial:operations ?C,C,C=**num2**:number !origin:location !conn1:ConIdent
　　　line(s):[389,217]
--> n !num1:number !conreq:operations !num2:number !origin:location !ocs(Insert(num3,
　　　{})):feature_data !conn1:ConIdent line(s): [364,269,227]
--> n !**num2**:number !**ring**:operations !dest:location !conn1:ConIdent line(s): [377,286,242]
--> f !conn1:ConIdent !**call_forward**:feature !Insert(num3, {}):number_set line(s): [412,254]
--> n !**num2**:number !**detect_forward**:operations !**num3**:number !dest:location
　　　!conn1:ConIdent line(s): [353,295,255]
--> n !num3:number !ring:operations !dest:location !conn1:ConIdent line(s): [377,286,242]
--> n !num3:number ! connect:operations ! dest:location !conn1:ConIdent lines:[377,289,247]
--> v !**violation**:error_msg !OCS:feature !ring:operations !conn1:ConIdent !num3:number
　　　line(s): [378]
End of trace.

The trace indicates a path that goes through phone *num2*, and a call forward operation to phone *num3*.

This experience shows that with this kind of tool there needs to be some search strategy defined by the analyst, but this strategy can be defined by the characteristics of one feature independently of the others. As a futher detail, one can note that in our search strategy a call forward feature to phone *num3* for phone *num2* was set up.Happily, LOTOS allows more generality than this. A variable can be used instead of a constant by utilizing the *distributed choice* construct. In this way, one can specify that a given phone instance has a call forward feature on, without specifying explicitly to which number. A tool including a *narrower* will try different values chosen from those declared in the ADT and eventually will find a solution. The following specification ca be used, instead of the previous one:

```
(
    (
        phone[u,n,f](num1,fwd(none),ocs(Insert(num3,{})) )
        | | |
        (choice X:number [] phone[u,n,f](num2,fwd(X),ocs({})) )
        | | |
        phone[u,n,f](num3,fwd(none),ocs({}))
    )

  |[n]|

    network[n]
)

  | |

verify_intentions[f,u,n]
```

The variable X stands for an unspecified forward number in expression:

(choice X:number [] phone[u,n,f](num2,fwd(X),ocs({})))

The results of the goal oriented search are identical to the previous case.

5. Conclusions

A method for detecting intention violations in the specification of telephony systems with features has been presented. The method can be effectively carried out by using the LOTOS language and associated tools.

Abstract data types are well suited for the verification of intentions because they are apt to specify constraints in an implementation-independent way and they can be used in a variety of language contexts. Specification of intentions is an important step toward feature interaction detection but for actual detection, some type of execution must be performed. We have seen how the goal-oriented execution method can help focussing execution and thus reducing combinatorial explosion. However for this to happen a lot of information must be provided to the tool. More research needs to be done in the field of static inspection of specifications to determine search strategies to reduce as much as possible the scope of execution techniques.

Acknowledgments. We are grateful for support of the Telecommunications Research Institute of Ontario, Bell-Northern Research, and the Natural Science and Engineering Research Council. Yow-Jian Lin of Bellcore asked us the 'right questions' that led to this work. We should thank Antoine Bonavita for having helped us with the goal-oriented tool.

References

[BE93] E. Brinksma and H. Eertink, Goal-Driven LOTOS Execution. In: A. Danthine, G. Leduc, and P. Wolper (eds). *Protocol Specification, Testing and Verification, XIII.* North-Holland, 1993, 45-60.

[BW94] W. Bouma and H. Velthuijsen. Introduction to the book *Feature Interactions in Telecommunications Systems,* IOS Press, 1994, vii-xiv.

[BA94] K.H. Braithwaite and J.M. Atlee. Towards automated detection of feature interactions. In: L.G. Bouma and H. Velthuijsen, *Feature Interactions in Telecommunications Systems,* IOS Press, 1994, 36-59.

[CGL94] E.J.Cameron, N.D.Griffeth, Y.-J. Lin, M.E.Nilson, W.K.Schnure and H. Velthuijsen. A feature interaction benchmark for IN and Beyond.In: L.G. Bouma and H. Velthuijsen, *Feature Interactions in Telecommunications Systems,* IOS Press, 1994,1-23.

[DB78] A. Danthine and J. Bremer. Modeling and Verification of End-to-End Transport Protocols. Computer Networks, 2 (1978), 381-395.

[DN94] O.C. Dahl and E. Najm. Specification and Detection of IN Service Interference Using LOTOS. In: R.L. Tenney, P.D. Amer, and M.U. Uyar (eds) *Formal Description Techniques, VI.* North-Holland, 1994, 53-69.

[FLS91] Faci, M., Logrippo, L., and Stépien, B. Formal Specification of Telephone Systems in LOTOS: The Constraint-Oriented Approach. *Computer Networks and ISDN Systems 21* (1991) 53-67.

[GV94] N.D.Griffeth and H. Velthuijsen.The negotiating agents approach to runtime feature interaction resolution. In: L.G. Bouma and H. Velthuijsen, *Feature Interactions in Telecommunications Systems,* IOS Press, 1994, 217-235.

[HLS93] M. Haj-Hussein, L. Logrippo, and J. Sincennes. Goal-oriented Execution of LOTOS Specifications. In: M. Diaz and R. Groz (Eds.) Formal Description Techniques, V. North-Holland, 1993, 311-327.

[Hol85] G.J. Holzmann. Backward Symbolic Execution of Protocols. In: Y.Yemini, R. Strom,

1985, 19-27.

[LL94] F.J.Lin and Y.-J. Lin. A building block approach to detecting and resolving feature interactions. In: L.G. Bouma and H. Velthuijsen, *Feature Interactions in Telecommunications Systems*, IOS Press, 1994,86-119.

[SL95] B. Stepien and L.Logrippo. Feature interaction detection by using backward reasoning with LOTOS. In: S.T. Vuong and S.T. Chanson. *Protocol Specification, Testing and Verification XIV*. Chapman & Hall, 1995, 71-86.

[VSV91]C. A. Vissers, G. Scollo, M. van Sinderen, and E. Brinksma, Specification Styles in Distributed Systems Design and Verification, Theoretical Computer Science 89, 1991, 179-206.

[Zav93] P. Zave. Feature Interactions and Formal Specifications in Telecommunications. IEEE Computer, August 1993, 20-29.

Feature Interaction Detection and Resolution in the Delphi framework

Martin Boström and Måns Engstedt
ELLEMTEL Utvecklings AB
Box 1505 125 25 Älvsjö, Sweden
{euambo,euamet}@eua.ericsson.se[1]

Abstract. In this paper, we introduce the formal specification language Delphi and describe how it can be used for the specification of telecommunication services and features. We also show how feature interactions can be detected and resolved within the Delphi framework. A methodology for specifying features and for detecting and resolving feature interactions is given.The need for general extension mechanisms in a logic-based specification language like Delphi is discussed. A language mechanism for non-monotonic extensions of the rule part of a Delphi specification is proposed. Non-monotonic logics and reasoning is proposed as the semantic base for these generalized extension mechanisms.

1 Introduction

1.1 The Feature interaction problem

Feature interaction is a well known problem in the telecommunication domain [1],[5]. With an increasing number of features with increasing complexity, feature interaction will be a major problem in telecommunication systems in the near future. Features should be specified in a fashion enabling additions of new features in a structured and controlled way. If features are described as modules in a formal and executable specification language using a proper methodology, automated tools can be used to detect some feature interactions. The detected feature interactions shall also be resolved in a structured and formal way.

---

1. This work is partially based on material developed in the RACE project SCORE [1], [3].

1.2 Reading instructions

It should be possible to follow the contents of this paper without any previous knowledge of the Delphi language and tools[2][2]. We start by giving a a specification example to introduce both the Delphi language and the different techniques for feature specifications. We also show how feature interaction is detected and resolved in the Delphi framework using the SafeGuard tool. A methodology to specify features and to detect and resolve feature interaction is also given. Finally some conclusions and proposals of future work are given.

1.3 Terminology

The used concepts are listed below. Their definitions agree with the definitions in [4], [5].

- A *service*, e.g. POTS, provides functionality that can be used on its own.

- A *(service) feature*, e.g. Call Waiting, provides added functionality to an existing service or feature. A feature is used together with a service and can not be used on its own.

- A *service interaction* occurs when one service affects the behaviour of another service.

- A *feature interaction* appears when one feature affects the behaviour of another feature.

We also introduce the following concepts:

- An *interaction feature* is a feature used to resolve some sort of feature interaction.

- A *joint specification* consists of a service specification, a set of feature specifications and possibly a set of corresponding interaction feature specifications.

Note that a basic service and a feature that interact is not a feature interaction according to the above definitions. However, the same resolution mechanisms that are used for feature interaction could be used also for this case.

2 Specifying features with Delphi

2.1 The Delphi language, an overview

The Delphi language is primarily intended to be used in the early phases of the service creation process in order to produce an implementation independent service specification, resulting from functional requirements on the service or the feature.

A Delphi specification consists of a part describing the static part of the modelled system, *the conceptual model (CM)*, and a part modelling the dynamics of the system, the *rules*.

The conceptual model describes which *entity types* that exist in the system, their static properties and their relationships. The conceptual model is thus a *schema* over entity types. From this schema, it is possible to create instantiations that will be sets of *entities* or *instances*. The static part of the specification determines the state space of the specified

---

2. The Delphi language and tools are under development at Ellemtel Utvecklings AB.

system. The conceptual model is given a semantics by translating it into a variant of first-order predicate logic.

The set of rules defines the state transition function of the system. A rule consists of a stimulus, conditions and conclusions. The informal semantics is that if the given stimulus occurs when the condition is true, the conclusion will hold in the next system state.

A Delphi entity type can have *attributes* and can include declarations of propositions, that may hold for the entity type. Between entity types, *relations* may be specified. In the conceptual model it is also possible to define static constraints (*invariants*), so that only certain combinations of propositions may be true for an entity type or certain relations may only exist when certain conditions hold.

A *module*, as used in this document, is a Delphi specification of a service or a feature. A module consists of a set of rules and a CM with a set of invariants describing the service or feature. Modules can be composed to become composite modules. The current Delphi language and tool set have no explicit support for modules.

2.2 The Delphi tools, a very short overview

The Delphi tools [2] consist of a graphical specification editor, a simulator, a theorem prover, the SafeGuard tool, a Use Case tool and a Test Case generator. Only the simulator and the SafeGuard tool are described and used here.

2.3 A Delphi CM prepared for feature extensions

A Delphi specification of POTS is shown in figure 1.

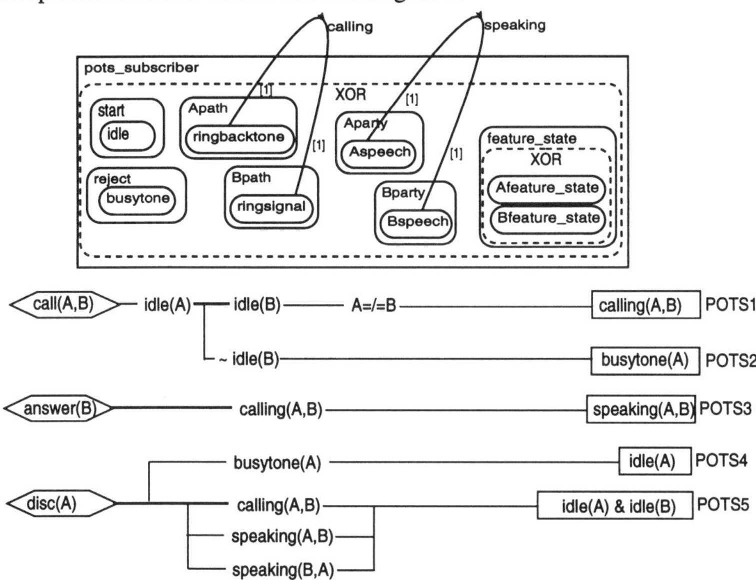

Figure 1. The POTS basic service (CM and rules)

The POTS service consists of a declaration of the entity type *pots_subscriber* and a set of rules, giving an abstract model of the external actions possible for a POTS user. The specification models the POTS application as a system consisting of instances of the *pots_subscriber* entity type. The entity type in the specification has a number of propositions or state declarations, e.g *start, idle* and *Aparty*, and two relations *calling* and *speaking*. The entity type uses nested propositions. Nesting one proposition inside another, means that the inner proposition logically implies the outer one or can be viewed as a substate declaration of the outer state declaration. The propositions on the outermost level are in this example mutually exclusive, i.e exactly one of them must be true at a time. This is indicated by using the XOR connective on the set of states. The language also supports other kinds of connectives, such as AND and OR, with the obvious semantics.

The relations in this example have been equipped with *cardinality constraints*. For instance, the 1:1 cardinality constraint on the *calling* relation in the figure implies that if a pots_subscriber entity has the proposition *ringbacktone* true, there must exist one instance of a pots_subscriber having the proposition *ringsignal* true, and the relation *calling(X, Y)* must hold between these two entities.

The *feature_state* proposition has been added to enable later additions of features to the CM. This is a mechanism to override the default behaviour specified for POTS. The *extension proposition* is false by default in the POTS service, but may become true when features later are added. A feature may then specialize this feature proposition by adding a local state declaration and an implication from the local declaration to the generic feature proposition declaration.

2.4 Delphi rules prepared for feature extensions

There are five rules in the POTS service, *POTS1, ..., POTS5*, each one consisting of (from left to right in the graphical rule) a stimulus, conditions and conclusions. The rules are described according to the graphical syntax of Delphi[3].

The first *rule graph* consists of two *rule branches*, with the names *POTS1* and *POTS2*, that share the same stimulus, *call(A,B)*. They also share the condition *idle(A)* but also have some conditions that are specific for each rule branch. POTS1 finally contains the conclusion *calling(A,B)*. A line between two conditions in the graph is interpreted as a logical conjunction (e.g. *idle(A) AND idle(B)* in POTS1). Two alternative paths between two points in a graph is interpreted as a logical disjunction of the corresponding conditions (e.g. *calling(A,B) OR speaking(A,B) OR speaking(B,A)* in POTS5). Each rule has a name that is written to the right of the graphical rule[4]. Several rule branches may be triggered at the same point in time by the same stimulus.

2.5 Two proposed extension mechanisms for rules

A service feature might add completely new rules that are independent of the rules of the basic service or it might add rules that has a relation to the original rules. We propose an

---

3. There is also a textual Delphi syntax not used here.

4. Rule names is a proposed extension to the current Delphi syntax.

enhancement of the present graphical rule syntax supporting modularization of services and features.

In figure 2, a rule graph S1 of the basic service is extended with 2 new rule branches, F1 and F2, that represent new service features that are added.

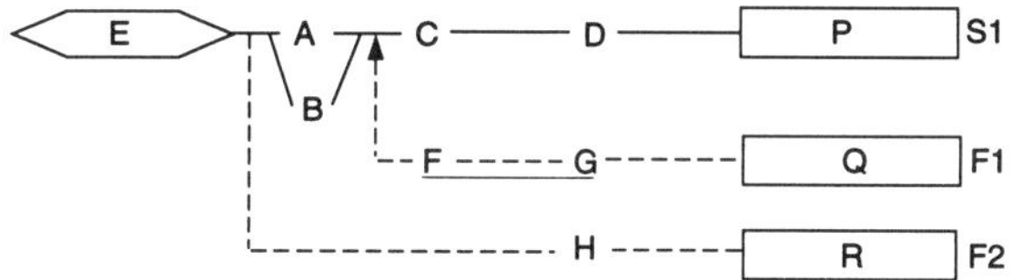

Figure 2. Extensions of a rule graph

Two types of new branches can be added to a rule graph. The first one, denoted with a dotted line with no arrow, is used to branch from any point in a rule graph. This is used for *monotonic* feature branches. Monotonicity in this context means that a base branch is not directly affected by an extension branch, nor does the base branch affect the extension branch for a given stimuli. The branch F2 in figure 2 is monotonic and is really the same type of branch that was used in the basic service rule graph but with a different graphical appearance to clearly separate it from the branches of the basic service and to support modularization of rules.

The second type of branch is denoted with a dotted line with an arrow, which indicates that this feature branch affects the base graph after the branching point, i.e. there is an interaction between the branches. The branch F1 interacts with the branch S1 in figure 2. The extension is *non-monotonic*, in contrast to the first type, in the sense that the new branch will restrain the conclusion of the base branch. This type of branch might also be added at any point in an existing rule graph.

The effect of the extension branch on the base branch, is determined by a *branching expression* denoted as an underlined condition in the non-monotonic branch. The branching expression can be anything from a Delphi literal to the complete formula on the branch between the branching point and the conclusion. The branching expression is implicitly conjoined negated to the base branch after the branching point. A branching expression that is not underlined means that all conditions of the branch is used as the branching expression. In figure 2 the branch F1 has the branching expression $F\ AND\ G^5$.

Conjoining a negated branching expression to a rule branch is a way of strengthening the conditions of that rule branch. Several non-monotonic branches with a number of branching expressions conjoined to a rule branch B will conjoin the conjunction of the negation of each of the branching expressions to the branch B.

The addition of a non-monotonic branch can be interpreted as a form of non-monotonic reasoning [6]. The non-monotonic rule branch can be viewed as an exceptional case, overriding the normal (default) case. This may also be viewed as a mechanism to make branches mutually exclusive and to give *priority* to an exceptional case over a normal case. If the feature branch is triggered then the updated branch, the normal case, will not be triggered and vice versa.

---

5. The underlining of the branching expression is in this case redundant.

In figure 2, the S1 branch will implicitly have the negation of the branching expression added, i.e. *NOT (F AND G)* which is equivalent to *(NOT F) OR (NOT G)*. The updated branch S1 is described in figure 3.

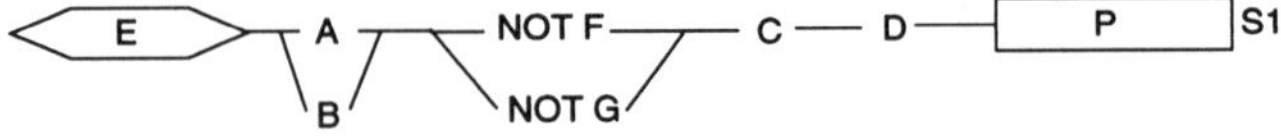

Figure 3. Rule branch updated with negated branching expressions

Non-monotonic branches can be added at any number of places recursively, which allows service features to be added to service features in any number of steps. Two non-monotonic rule branches B1 and B2, added at the same point of another rule branch, might interact with each other. The interaction can be resolved by giving priority to one of the branches over the other either with a graphical or textual notation. Non-resolved interactions can always be automatically discovered by the SafeGuard tool due to a proved inconsistency (see below).

The described branching mechanisms thus support

- graphical separation of rule branches in a clearly visible, declarative and modular way

- a controlled way of resolving undesired interactions between rule branches which cause inconsistencies without having to make explicit changes in the existing rule base

2.6 A textual notation for priority

An explicit textual priority relation > expresses priority between rule branches. *A(BE) > B* means that rule branch *A* with a branching expression *BE* has priority over rule branch *B*, i.e. if *BE* is true then *A is* triggered, otherwise (if *BE* is false) *B* is triggered.

The priority relation in figure 2 between the branches F1 and S1 could instead be stated textually as: *F1(F AND G) > S1*.

2.7 Interaction in the Delphi framework

Given a joint specification of a basic service and some features, the specification is free from feature interaction if

- the joint specification is consistent.

- the properties specified for the individual features are fulfilled by the joint specification.
   This is to some extent a simplification since there might be interactions, according to our earlier definition, even with a consistent joint specification, but here we only look at interactions showing up as inconsistencies.

2.8 The SafeGuard tool and feature interaction detection

The SafeGuard tool makes it possible to prove both consistency and properties given in a simple property language. Presently only safety properties like *unless, stable, invariant* and *sound rules* (see below) are possible to prove in SafeGuard. A proof is done by translating the specification to a state machine representation in propositional logic and by using an efficient theorem prover.

The SafeGuard tool is used to detect feature interactions in a joint Delphi specification. SafeGuard will automatically discover an inconsistency in a joint specification for an instantiation with a given domain size. SafeGuard then gives a warning and a diagnostic printout showing a possible inconsistent state or a transition with corresponding conclusions leading to an inconsistent state. The Delphi simulator can also be used to discover inconsistencies but in a more indirect way than SafeGuard. The simulator is on the other hand more useful to detect an incomplete specification.

The proofs in SafeGuard are limited to instantiated specifications, i.e specifications for a given number of subscribers. Freedom from interaction with a given instantiation does not mean that no interaction occurs with a different number of subscribers. It seems however that most interactions, due to the structure of the joint specifications, can be captured by analyzing rather small instantiations, i.e. increasing the domain size above a threshold value N will not cause any new cases of interaction. One problem is to decide N which will depend on the set of involved features.

2.9 Inconsistent Delphi specifications

A given joint specification can be inconsistent in two ways. First of all, the conceptual model can be inconsistent. Given a consistent conceptual model it is also possible that the rules together with the conceptual model are inconsistent. In order to prove consistency of a given specification we must therefore prove both the consistency of the conceptual model as well as the soundness of the rules.

A failure to prove the consistency of the conceptual model and a failure to prove the soundness of rules are both automatically detected with SafeGuard. The resolution of the inconsistency is another matter.

2.9.1 Inconsistent conceptual model

A conceptual model is formally represented as a set of Delphi logical formulas known as the static theory [2]. A given specification, with a static theory (conceptual model) T, has a *consistent conceptual model* if and only if it is not the case that $T \vdash \perp$.

A proof that the actual conceptual model is consistent (or inconsistent) is the first step towards detection of feature interaction.

2.9.2 Unsound rules

A rule in a specification is said to be *unsound* if it is possible to trigger the rule in a consistent state and the resulting state is the error state, $\otimes$ [2]. The error state is any inconsistent state, i.e a state where both the negation of a proposition and its negation are concluded.

We believe that unsound rules is the most common sign of feature interaction. A typical example is when two features trigger different branches of the rule for the same stimulus and the different drawn conclusions together are inconsistent with the conceptual model.

2.10 An isA-hierarchy of feature entity types

In this section we present how the *isA* relation may be used to extend the conceptual model of a service with features.

In our specification approach, the POTS specification consists of just one entity type. For each feature we specify, we define an entity type that is related to the basic service entity type by the *isA*-relation. These feature entity types inherit attributes etc. from the *pots_subscriber* entity type but may declare propositions, attributes, relations and static constraints of their own.

A Delphi specification of the Call Forwarding on Busy feature (CFB) in relation to POTS is shown in figure 4. CFB is only allowed to transfer calls in one step in this version of the feature.

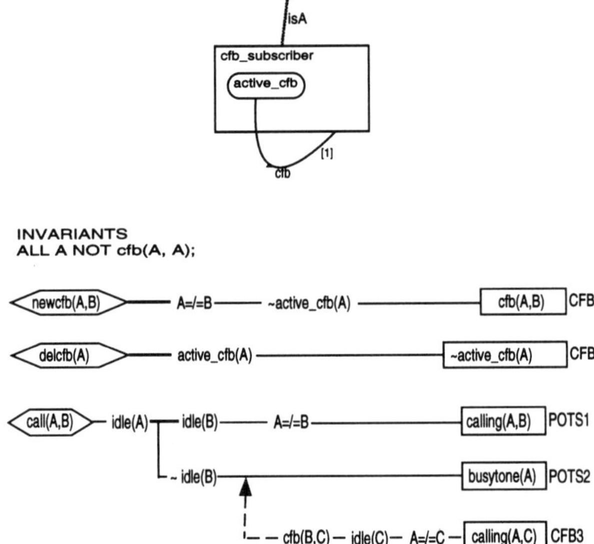

Figure 4. The Call Forwarding on Busy feature

All states, attributes and relations that have a connection to the CFB feature are gathered in the entity type *cfb_subscriber* to clearly separate them from the basic service and the other features. It is, however, only a subscriber that has the proposition *active_cfb* true that really subscribes to the CFB service. A subscriber might be of entity type cfb_subscriber without having the CFB feature active.

Each feature will have its own entity type when the features are specified in our examples, but this is certainly not the only way to model a set of features. The composite CM of a set of features could also have been modelled with only one entity type, but this would have made it very hard to visually separate the different features.

The entity type *cfb_subscriber* has an *isA*-relation to the POTS subscriber which means that the *cfb_subscriber* inherits propositions, invariants etc. from the *pots_subscriber*. The relation *cfb*, with a corresponding cardinality constraint, shows that a *cfb_subscriber* can have a cfb-relation to only one other subscriber. This other subscriber does *not* need to subscribe to the CFB feature. A textual invariant expresses that a cfb_subscriber can't have a cfb-relation to itself.

2.11 Delphi feature rules

The set of rules for each feature is in the example divided in two parts. The first part specifies *activation* and *deactivation* of the feature in question. The second part specifies the rules that could trigger when the feature has been activated, the *traffic handling cases*.

The rules *CFB1* and *CFB2* are activation and deactivation rules. The rule *CFB3* is a traffic handling rule interacting with POTS via a non-monotonic rule branch with an underlined branching expression *cfb(B,C) AND idle(C)* that is added negated to the affected POTS rule branch *POTS2*, i.e. *NOT SOME C (cfb(B,C) AND idle(C))* is implicitly added to *POTS2*. Since the variable *C* is free in the rule, it has to be existentially quantified in the negated branching expression. This mechanism gives priority to CFB over POTS in the case where a busy subscriber *B* has calls forwarded to an idle subscriber *C*.

Note that calls that are forwarded to a busy subscriber will be handled by *POTS2*. Without the priority mechanism, *POTS2* could be triggered by the same stimulus as *CFB3* which is an undesired, and even inconsistent, result. This interaction, which would be automatically discovered by SafeGuard, is conceptually not considered as a feature interaction since POTS is a service and not a feature.

A Delphi specification of CW in relation to POTS is shown in figure 5.

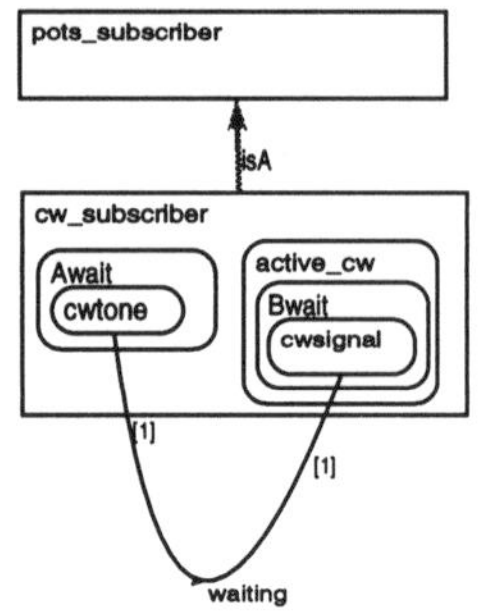

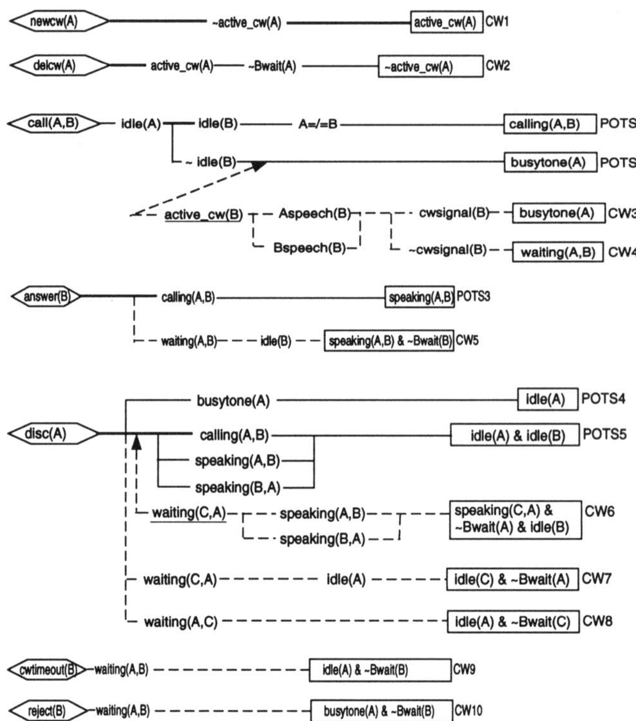

Figure 5. The Call Waiting Feature

The POTS rule branches *POTS2* and *POTS5* in this example are affected by non-monotonic rule branches of CW which in these cases give priority to CW over POTS. It should be noted that *POTS2* was affected by CFB in similar way.

2.12 Interaction detection, Call Waiting-Call Forwarding on Busy

If a subscriber has both CFB and CW active, there will be a conflict in handling calls that are originated towards the subscriber when he is busy. If the *CW4* rule triggers the conclusion *waiting(A,B)* is drawn. If the forwarded-to subscriber is idle then the *CFB3* rule draws the conclusion *calling(A,C)*. The two conclusions can't be valid simultaneously since the A-subscriber can't be a member of both relations according to the specified joint CM described later. According to the CM, subscriber A can't be in the mutually exclusive states *ringbacktone* and *feature_state* simultaneously. If the joint CFB and CW specification is run through the SafeGuard tool this will automatically show up as a case of *unsound rules (inconsistent conclusions)* for a certain transition, i.e. the feature interaction is automatically detected with SafeGuard.

2.13 Interaction resolution, Call Waiting-Call Forwarding on Busy

The interaction between the features CFB and CW is in this case resolved by assigning priority to CFB over CW[6]. The priorities resolving the interaction are here expressed as

CFB3(NOT SOME C (cfb(B,C) AND idle(C))) > CW3
CFB3(NOT SOME C (cfb(B,C) AND idle(C))) > CW4

These two priority relations form an *interaction feature*. The resulting joint rule specification, including the POTS rule branches, with expanded branching conditions is shown in figure 6.

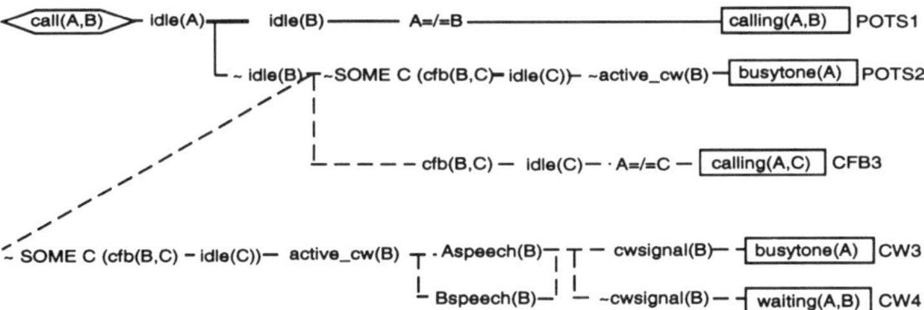

Figure 6. The resolved feature interaction between CFB and CW

Figure 6 shows what the rules would look like if the two branching expressions *cfb(B,C) AND idle(C)* in rule branch *CFB3* and *active_cw(B)* in rule branches *CW3* and *CW4* are added negated to the rule branch *POTS2* and if the branching expression *cfb(B,C) AND idle(C)* in rule branch *CFB3* is added negated to the rule branches *CW3* and *CW4*.

All the above rule branches are mutually exclusive, i.e. no inconsistent conclusions are drawn. *CFB* is given priority over *CW*. This means that if a subscriber has both *CFB* and *CW* active, calls that arrive when the B-subscriber is busy and when the subscriber *C* is idle, will be forwarded to *C* and no waiting call will be set up.

---

6. There are of course other ways of resolving this interaction.

2.14 A joint feature specification

In figure 7, a joint CM specification of POTS and a set of features, Calling Line Identification Presentation (CLIP), Call Forwarding on Busy (CFB), Call Waiting (CW) and Call Hold (CH) is shown. The CM specification is a merge or a composition of the CMs of the basic service and all the feature modules. In the case where the individual CMs are consistent, composition of the CMs is just an addition of all the isA-relations of each feature entity type to the POTS entity type. This is possible since each feature only adds a specialization of the pots_subscriber by using an isA-relation without changing the pots_subscriber entity type. The addition of a new feature entity type may of course cause an inconsistent conceptual model.

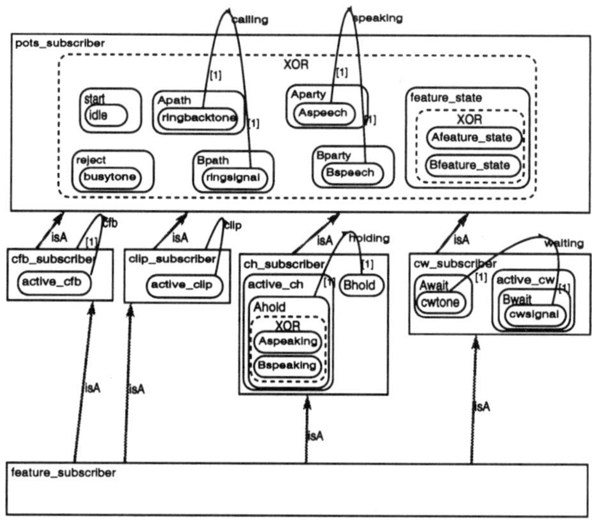

Figure 7. The joint conceptual model specification

At the bottom of the *isA*-hierarchy, the entity type *feature_subscriber* is added. This entity type is used for validation purposes. When running SafeGuard, an extra invariant can state that all entities are of the type *feature_subscriber* that can have any combinations of features active. This entity type inherits all other feature entity types and guarantees that all possible combinations of active features are validated when SafeGuard is used for validation of a set of features.

A joint invariant specification is just the union of the individual invariant specifications for POTS and each feature.

The joint rule specification, not shown here, is the union of the updated individual sets of rules for POTS and each feature. The updating of individual rules are caused by textual priority relations or individual non-monotonic rules.

Composition of feature modules might in the general case be much more complicated if the joint CM and the joint rules are inconsistent. General resolution mechanisms, like the priority mechanism, are then needed when the features are composed.

3 A specification and validation methodology

A methodology used for feature specification and feature interaction detection and resolution has been developed. The *SafeGuard* tool is used to detect inconsistencies in any type of specification since some types of feature interactions between a set of features show up as inconsistencies in the SafeGuard tool. Note that inconsistencies of course also can be caused by what is not defined as feature interactions, e.g. an inconsistency within a service specification or within a service and a feature specification. In addition, the *Delphi simulator* is used to simulate specified use cases, to validate that the requirements are correctly and completely captured. When a feature interaction is discovered, it is resolved by using a resolution mechanism.

The basic service, the features and the interaction features in the example were structured in three levels as described in figure 8.

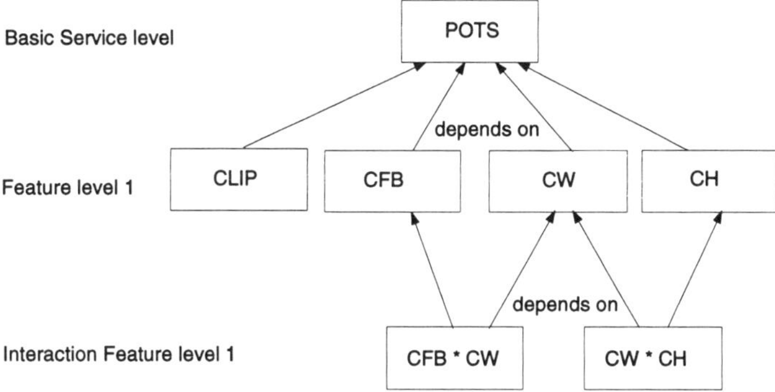

Figure 8. A service and feature hierarchy

The first level is the *(basic) service level* that isn't depending on any other service or features. This level only contains the POTS service.

At the second level, the *feature level*, the features, here CLIP, CFB, CW and CH, are specified in relation to the service level, here consisting of the POTS service. This is done by using the mentioned techniques such as using an *isA*-hierarchy of entity types including a multiple feature entity type, using extension properties in the Delphi formulas and using non-monotonic rule extensions. There can also be higher order features at higher feature levels that are depending on features at lower levels.

Above the service level and the feature levels comes finally the *feature interaction level* with the interaction features, here CFB*CW and CW*CH. CFB*CW denotes an interaction feature used to resolve a feature interaction between CFB and CW. Interaction is resolved by using the priority mechanism. For interaction features, as well as for pure features, there can be higher order interaction features although such interaction features were not needed in our example.

The following methodology was developed as a result of the used service and feature hierarchy:

1. *Basic service specification.* First the basic service is specified as a separate module in a way that enables extensions and that uses the proposed specification structure.

2. *Basic service validation.* The basic service is first validated using the SafeGuard tool. This guarantees a consistent specification. Simulations are then performed to validate that the specification is complete and that the requirements have been correctly captured. If the specification is inconsistent, incomplete or incorrect, then the Basic service specification phase must be entered again.

3. *Single feature specification.*The feature is specified in relation to the basic service and a (possibly empty) feature set that the feature is depending on (see figure 8), using the proposed specification structure and extension mechanisms. A new feature entity type is created in the isA-hierarchy, new invariants and rules are added. Each feature is specified as a separate module consisting of a set of rules and a CM with isA-relations to other services or features.

4. *Single feature validation.* The feature and the basic service (plus features on lower levels) are validated together using SafeGuard and the simulator. If the feature specification is inconsistent, incomplete or incorrect, a change should preferably only be done in the new feature, which is at the top level in the feature hierarchy, and not in the basic service or in lower level features. Changes at a lower level should only be done in exceptional cases.

5. *The previous steps are repeated for each new feature to specify.* If each feature is validated with the basic service, then each feature is at least correctly and completely specified when looked on in absence of other features.

6. *Multiple feature validation.* A validation is then performed on the joint specification, i.e. the basic service and all features. At this stage, according to our definition, feature interactions can appear as inconsistencies. If a feature interaction is detected, one or several interaction features are specified, otherwise the set of features is considered to be validated.

7. *Single/multiple interaction feature specification.* An interaction feature(s) is specified using the proposed mechanisms, such as the non-monotonic priority mechanism. One interaction feature could resolve an interaction between a pair of features or it could resolve all interactions between a set of features. The first case, *case-by-case approach*, is used in the example.

8. *Single/multiple interaction feature validation.* The basic service, the interacting features and the interaction feature(s) are validated together. If this joint specification is inconsistent, preferably only the interaction features should be changed. Changes at a lower level should only be done in exceptional cases.

The basic principle for specification is thus that specifications are done first for the service, then for features on level 1, then for features at level 2 etc., and finally for the interaction features. Validation is done in the same order. Inconsistencies and interactions should preferably never be resolved at a lower level than the one being validated at the moment.

4　Conclusions

Delphi is a very suitable framework for specifying and analyzing services and features on a high abstraction level. As can be judged from initial case studies [3], the proposed mechanisms for service and feature specification seem to work with satisfying result. The mechanisms together with the used methodology, the Delphi language and the tools provide a good starting point for feature specification and feature interaction detection and resolution.

The analyses and the diagnostics from SafeGuard when used on a set of interacting features should however be improved in order to facilitate the feature interaction resolution step.

If we do not want to change existing modules when new features are added, then some sort of non-monotonic logic and reasoning [6] is needed to resolve feature interactions. A feature must be able to restrain or override assumptions made in a base specification. This must be specified outside the base specification in the feature itself or in an interaction feature in order to avoid explicit dependencies from the base to the feature. Delphi of today does not have any such mechanisms, neither for the rules nor for the CM. We have proposed a mechanism for such non-monotonic extensions of the rules. For the CM we have discussed the approach of using extension propositions. There is obviously a need for more general extension mechanisms.

An example of conflicting cardinality constraints in the CM are shown in figure 9.

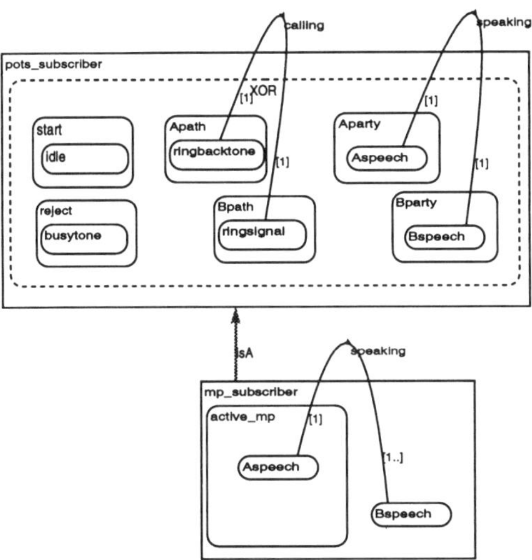

Figure 9. Conflicting constraints in the CM

A cardinality constraint in the pots_subscriber says that one POTS subscriber can have a *speaking*-relation to another POTS subscriber. A multiparty subscriber must relax this constraint to enable multiple speaking-relations to other subscribers. This means that the constraint for POTS should be replaced by a weaker constraint for the multiparty feature.

This is expressed in the isA-hierarchy by adding a relaxed speaking-relation cardinality in a multiparty entity type as described in figure 9.

These two constraints of the speaking-relation in figure 9 can result in a conflict. The semantics of such conflicting constraints in the CM could be that a constraint further down in the isA-hierarchy always overrides a constraint further up in the hierarchy. In this case the Multiparty constraint overrides the POTS constraint. This is one mechanism to automatically resolve inconsistencies in the CM.

A non-monotonic priority mechanism similar to the one used for rules is thus needed also for formulas in the conceptual model as well as for the textual invariants.

5 Future work

The current Delphi language has no support for modularization of specifications. Large systems must be handled in a modular way also on the specification level, both from a system management and reuse point of view.

The Delphi language should be enhanced with a module definition and in addition to that enhanced with general non-monotonic reasoning mechanisms for the conceptual model and the rules.

Today it is possible to use the SafeGuard tool to detect feature interaction appearing as inconsistencies. Tool support to detect consistent interactions should be included in the future Delphi tool set.

References

[1]   SCORE-Methods and Tools. *Report on Methods and Tools for Service Creation*. SCORE Deliverable D204, vol. 1, 1994.

[2]   M. Boström, O. Öhlund. *The Delphi Language and Tools*. SCORE deliverable d202p1e1, 1994.

[3]   M. Boström, M, Engstedt, O. Öhlund. *Feature Specification, Interaction Detection and Interaction Resolution using Delphi, a Case Study*. SCORE deliverable d202p1e2, 1994.

[4]   E.J. Cameron, N. Griffeth, Y. Lin, and H. Velthuijsen. "Definitions of Services, Features, and Feature Interactions", December 1992. Bellcore Memorandum for Discussion, presented at the 1:st International Workshop on Feature Interactions in Telecommunications Software Systems.

[5]   E.J. Cameron, H. Velthuijsen. *Feature interaction in Telecommunication systems*. IEEE Communications Magazine, 31(8):18-23, August 1993.

[6]   D. McDermott, J. Doyle. *Non-Monotonic Logic* I. Artificial Intelligence, 13(1-2), pages 41-72, 1980.

Service Validation and Testing

Bryce Kelly, Michael Crowther, John King, Ronnie Masson and Jacques DeLapeyre

BT Laboratories, Martlesham Heath, Ipswich, UK
bryce@geatland.bt.co.uk

Abstract. In the context of evaluating IN service implementations from vendors and BT , we have developed and applied structured modelling tools to detect and resolve feature interactions during service validation and testing. Very successful and promising results have been obtained which supplement and enhance the traditional testing methods used to evaluate new services from vendors.

We describe practical results from two types of models built to investigate feature interaction. The results have forced implementation changes by manufacturers before the service could be offered by BT.   The first model, a Centrex network model, supports several nodes and its architecture includes techniques for detecting feature interactions. Its central feature is an interaction manager , which not only allows detection, but also allows different sequencing of features to test different service invocation orders. This dynamic flexibility of controllable non-determinism provides a very powerful means of resolving interactions, compared to deterministic models or actual implementations.

The second model , a service plane model, is based on a simplified abstract view of the telephony service together with the features which enhance it. The network is viewed as a single node of call control and resources, with dynamic creation of associations between instances of user parameters, basic call and features. Mechanisms have been developed to allow features to be independently created and added freely to the model. This allows decisions on feature interaction and resolution to be based on the actual results of service operation and animation rather than on preconceived rules on how services might be expected to interact.

The underlying technology behind these projects is the standard specification language SDL, with its powerful commercial tool support. This has allowed the accuracy of our modelling and the validity of the results to be independently assessed by feature specifiers [5]. The rigour of the initial specifications as well as the easily observed graphical animation capability of SDL has considerably eased that task. The potential for automated test generation using SDL specifications is a further benefit [6].

1. Introduction

The subjects of service validation and testing are rapidly gaining importance for service providers and network operators. The overall aim is to ensure that all existing and future services and features operate as specified and interact in a sensible, intended, manner.

Here we consider service validation to be the process of establishing that a service specification meets its requirements and interacts sensibly with other services at the specification level. Service testing is the process of establishing that an implementation of a service meets its specification and interacts sensibly with other service implementations.

The roles of service validation and testing may to some extent be dependent upon the type of platform used. Services specifications to be supported on switches and other dedicated platforms must be validated before being given to the manufacturer. The implementation of the services must then be tested before being connected to the network. However for completely generic service platforms with an advanced Service Creation Environment, as may be the case in an Intelligent Network (IN), the role of service testing is less clear. The need for service validation may indeed be greater. In practice we are likely to see a mix of these two types of platforms for some time to come and so we consider it important to study validation and testing. Indeed it is our belief that the methods involved in service validation and test generation are inextricably linked.

Probably the most important aspect of service validation is the detection and resolution of unintended feature interactions which lead to unintended behaviour of a service or feature.

The problem of unintended feature interaction already exists within the scope of current network services. These are usually dealt with on a case by case basis using manual investigation techniques. However with the rapidly increasing number of potentially interacting services and features being introduced, substantial importance is now being placed upon developing rigorous and automated techniques.

The concept of a multi paradigm approach to solve the feature interaction problem now seems to be becoming generally accepted on the basis that it will not be feasible to resolve all possible feature interactions at any one stage in the lifecycle or with any one technique. Additionally a 'divide and conquer' philosophy is proposed in [2] whereby the problem is split into a number of sub problems which can be dealt with using specific techniques. However we also believe that the well accepted principle of removing as many errors as possible at the beginning of a products lifecycle equally well applies to features and their interaction. This has led us to focus on techniques of avoidance and especially detection of interaction during the specification of features.

The subject of service testing is relatively easy when only considering services in isolation. Services themselves do not typically exhibit large degrees of complexity in their behaviour. Therefore it is fairly straightforward to derive the necessary set of tests required to establish a high degree of confidence in the implementation.

However as the number of services increases the number of potential tests can become unmanageable and uneconomic and so some degree of test selection is necessary. We have found that the service scenarios identified during service validation which display feature interaction should form an important part of a test suite for a set of interacting services. Additionally we believe that any ambiguity or unexpected complexity found during service validation on a model should be investigated on the real implementations. We will show later that this view proved justified in a real application.

When adopting rigorous design principles we might expect that feature interactions would either be removed by re-specification or catered for with some interaction resolution technique at the specification stage. In this case the scenarios are still likely to merit testing, especially if they had serious consequences.

A more practical approach to the problem which is being used in industry at the moment is to develop a model, lagging a short time behind the actual specification and implementation processes. The detected interaction scenarios are then directly applied to the testing process to establish exactly what the implementation provides in these potential problem areas. This is the approach we highlight in this paper.

The general approach for investigating service interworking and test generation described here is based upon modelling features using the ITU's standard specification language, SDL. Feature models are then used as a basis for analysing the behaviour of interacting features. In our work we have been concerned with modelling both switch based and non switch based, i.e. Intelligent Network, features, details of which are presented later. From past experience we believe that SDL is a suitable language for modelling services. SDL is very good for describing systems of an interactive nature, which is an important characteristic of the features that we wish to model. Additionally SDL could be viewed as a formalisation of 'flowcharts', which are commonly used at present when developing new features.

As this is a relatively new area of work the terminology used is not yet fully stable or universally agreed. There is no common agreement on the difference between a service and a feature. However the 'Implementors view' [1] seems to be useful and so will form part of the terms of reference of this paper. This view considers a feature as an increment of functionality added to the basic (usually telephony) service. In general unintended interaction in services is likely to be due to unintended feature interaction. Therefore throughout this paper the term feature interaction will be used to refer to any type of unintended interaction.

2. Methods overview

We consider two aspects of avoidance of feature interaction. The first is to use a rigorous technique for specifying the feature; in common with some others we have chosen SDL [3]. Although the use of a rigorous specification technique may seem rather obvious, its importance should not be underestimated. Our initial results show considerable potential benefits in this area as the precise specifications which can be obtained reduce the scope for misinterpretation of the feature behaviour and of interaction with other features. It is interesting to note that a significant number of feature interactions which have occurred within the network have been due to incomplete and imprecise specifications of features and how they interwork with others.

The approach assumes that detailed service specifications exist and that the network operator has access to them. Our models have been derived from a number of sources including BT specifications for Centrex and other core network services. The feature specifications in each of these typically exhibit some degree of incompleteness and imprecision. Modelling such specifications by using what is normally considered to be a precise language, i.e. SDL, is proving to be an interesting challenge.

There is an issue of accuracy of our models which can only be addressed by thorough review by feature specifiers. The models must therefore be readily understandable to these people. The graphical presentation of SDL is useful here as it helps to communicate an intuitive understanding of the specified behaviour.

The second aspect of avoidance requires analysis of groups of features. Our proposed method can be described by the following stages:

i. Develop a class of model sufficiently abstract so that the model design itself doesn't attempt to resolve feature interactions but has sufficient analytical power to detect potential interactions. Any specified feature interaction resolution techniques and data should be included in the model. We have developed two types of model which are discussed below.

ii. Model the environment of features. This represents the basic service which features are designed to augment together with mechanisms (such as trigger points) supporting the operation of features.

iii. Construct feature combinations of interest (scenarios) and model each feature in the context of the environment model.

In general it may be possible to analyse a single model containing all features of a service offering. However, a more effective approach will be to model groups of features which are important from a marketing viewpoint or which are thought likely to interact. Analysis of a number of these simpler feature scenarios is likely to prove more tractable, reducing the scope of multiple feature interactions and the number of behaviour traces to be derived and analysed. Nevertheless, scenario analysis is a major challenge when scaling up our interaction detection method.

iv. Explore the behaviour of each scenario using animation techniques, choosing model behaviour according to some coverage heuristics. We will be looking for effects indicating the potential for interaction between features.

A method for detecting interactions involves noting what the behaviour of each feature should be in combination with the basic call only, then comparing this with its behaviour when other features are introduced. The 'behaviour' is defined as the behaviour observable by the service users. This method results in much the same testing activities as those associated with real networks. Test stimuli are used to exercise the model and the resulting outputs are compared with those expected.

There are some practical problems with this black box testing approach. Interaction of features during analysis may lead to deadlock or non-deterministic choice in the model. These effects may not be directly observable by the users or may be difficult to observe due to complexity of the behaviour. We therefore need some idea of what the interaction mechanisms are in our models in order to detect these internally.

3. Models

There are a number of key issues which must be considered when developing service models. Probably the most important of these are the level of abstraction of the model and the approach to non-deterministic behaviour.

The appropriate level of abstraction should allow the detection of feature interactions whilst not overwhelming the analysis with unnecessary detail. The main abstraction factor in our models is the view of distribution of call control in the network. The most abstract view is to consider the call control and associated features for all service users to be provided from a single location. There is then no requirement to model any signalling between network elements. This is the approach taken for our service plane model. Less abstract models can be produced by distributing the call control and features between a number of network nodes. This requires a protocol to be modelled between the network nodes and is the approach taken by our network oriented models such as the Centrex network model.

When choosing a level of abstraction consideration needs to be given to the testing interfaces required for service testing. Clearly in all circumstances user access interfaces should be used for invoking and monitoring services under test. This provides a users view of the service behaviours. Both approaches described above support this. Less abstract models can uncover, for example, protocol behaviour related to services, which may be deemed important for testing. The Centrex network model could support these testing interfaces but the service plane model could not.

The way in which we are able to capture non-determinism in a model is also vitally important to establishing a direct relationship between the behaviour of the model and the behaviour of the original feature specification. Ambiguity, imprecision and incompleteness of the original specification can all be viewed as non-deterministic characteristics when trying to assign a precise meaning to it. It is therefore desirable to capture the non-deterministic properties of the feature in the developed model. For example, using the concurrency properties of SDL it is possible to model independent operation of multiple features. This is necessary as in practice many features are defined in isolation from each other. It is therefore unclear what their relationship is, including the order of invocation, in a multi feature environment.

The following sections present the two types of model we have built to investigate feature interaction. The first of these addresses Centrex network supplementary services. The second type of model addresses features at the Service Plane level of the IN conceptual model and is not dependent on any particular underlying network platform.

3.1 Centrex network model

The network oriented model described here was built to investigate interactions between Centrex supplementary services. Centrex provides private network services to users connected to a public network. The model uses a distributed view of call control similar to the view used by the service specifications from which it was developed.

In order to preserve model simplicity we have used a network model in which each node supports a single user and one of the network nodes becomes the focal point for interaction detection. This is termed the 'complete node' and has been designed so that its associated user may initiate or receive a basic call and any of the supplementary services being investigated. All other nodes within the network model are partially defined. A 'partial node' which is the originator of a supplementary service is modelled so that it may only originate that one service and handle its half of the basic call. A 'partial node' which terminates a supplementary service is modelled so that it may only terminate that one service and handle its half of the basic call. The model is illustrated in figure 1.

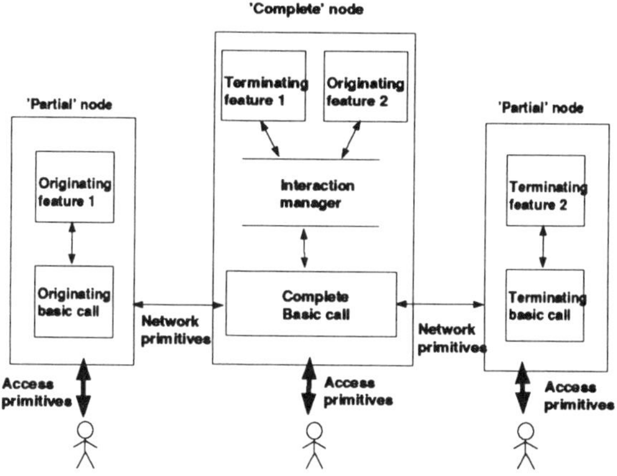

Figure 1, Centrex network model

Techniques for detecting interactions were also taken into account when developing the architecture of the model. Any interactions should be detected by observing the users' view of the behaviour of both their basic call and the supplementary services they have in operation (in this instance, we did not make use of the network signalling interfaces revealed by this relatively detailed model). The model also assists in the identification of causes of interactions via an interaction manager which intercepts messages to and from each supplementary service. The interaction manager also allows us to alter the sequences of messages to and from each supplementary service. This provides a degree of controllable non-determinism in the model, which brings considerably more flexibility than fully deterministic models. This is important as we are then able to test different service invocation orders in which interworking decisions have not been made in the service specifications.

The model was developed from existing procurement specifications and was used to identify potential problems with the Centrex implementation delivered to BT. Test sequences were derived for all the feasible combinations of features on the model. By simulating the test sequences on the model, we were able to note the results we would expect to see from performing the same tests on the implementation. Some of the model tests resulted in unexpected or ambiguous outcomes. By repeating the tests on the implementation and comparing the two sets of results, differences between feature operation requested by BT and delivered feature operation could be discovered and highlighted. The tests which led to "interesting" results during model testing were given priority.

The most useful result to date has been to identify a problem in the way that Call back and Diversion features interact in the Centrex switch software. The scenario concerned is where a calling user with the Call back feature calls a user with the Divert feature active. This is illustrated in figure 2.

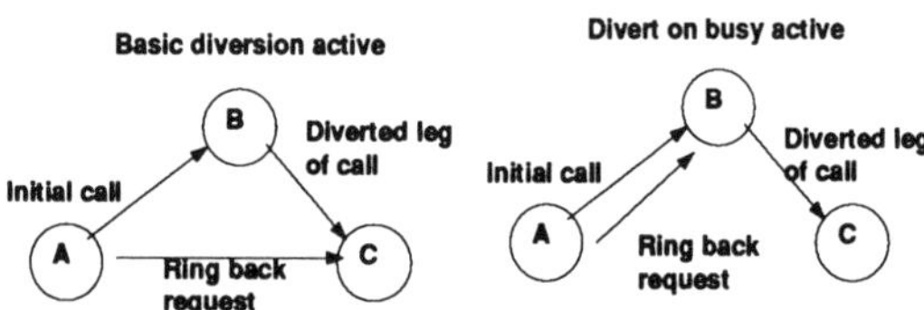

Figure 2, Call back request meets busy

The problem relates to the target of the Call back. It could be directed at the originally called user or at the user the call was diverted to. Our model, derived from BT's specifications calls back the "diverted to" user in all cases. Testing, based on our test scenarios, revealed that the behaviour of the supplied switch software depends on the kind of diversion encountered. For divert on busy it calls back the originally called user. The resolution of this situation is currently taking place.

3.2 Service Plane model

This model represents an abstract view of the telephony service and of features designed to augment it. This abstract view enables analysis of service features to focus on potential interactions which are implied by the feature specifications alone and to ignore complications introduced by implementation issues. A service plane model is useful in addressing interaction issues of IN based services, but can be applied to services based on any underlying service platform. The simplicity of the model also has the important advantage that it is more easily analysed using both manual and automated methods. The network is viewed as a single location of call control and of network resources. The model, shown in figure 3, has four types of entity which model the behaviour of telephone calls and the environment in which additional telephony service features operate.

Each *User Agent* entity encapsulates data which is specific to a particular user including the operational state of that user (idle, speech, etc.) and which features the user subscribes to. An instance of user agent is permanently associated with a particular user and so its data is valid over all calls the user may be involved in over a period of time.

Each *Basic call* entity is created and terminated as required and encapsulates the state of a particular call. It provides the basic function of controlling calls between users by ac-

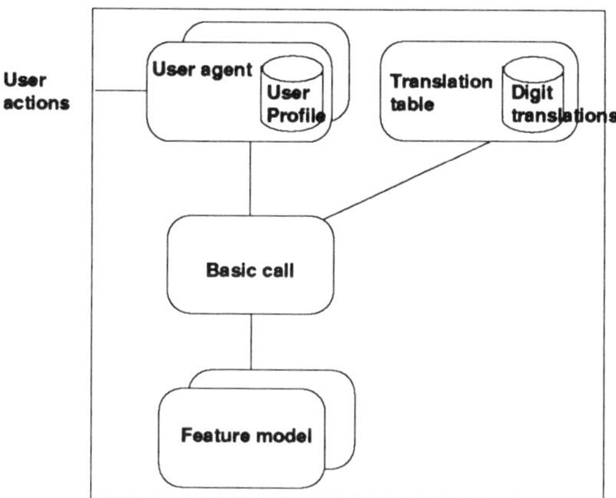

Figure 3, Service plane model

cepting user instructions for setting up and clearing the call. In the process, certain points in the call are identified at which additional service features may become active. The basic call is responsible for handling the invocation of these features and any effects they may have on the call state.

There is a single ***Translation Table*** entity which encapsulates network data which is not associated with a particular user, such as mappings for dialled digits. Instances of basic call or feature can request the translation table to perform mappings and to return the result.

Service features are modelled as independent entities which are created according to triggering information during the operation of a basic call instance. Features we have modelled so far include Diversion, Call waiting, Call back, Conference calling and Calling number delivery.

The operation of the model results in the dynamic creation and termination of associations between instances of user agent, basic call and feature. Different kinds of call configuration are possible. The model allows more than one feature to be actively associated with a basic call instance at one time. An example of this would be where a caller with the *Call back when free* feature calls a busy line equipped with the *Call waiting* feature. Furthermore, a feature associated with one basic call can change the global state of a different basic call. For example, the *Call back when free* feature requires that the previously set-up call inform the CBWF feature when it completes.

These complex call configurations can lead to two problems. Messages to users from a set of active features may be contradictory and requests for changes to the call state may be contradictory. The message sequence diagram shown in figure 4 shows a fragment of model behaviour, showing an interaction between the *Call Waiting* and *Divert on busy* features.

The call shown has been made to a busy user with both Call waiting and Divert on busy features. When the two feature models complete, they both request a change to the users to be connected by the call. The requests are different and not compatible, an interaction has been detected.

Mechanisms have been developed allowing features to be independently developed and freely added to the model. The basic idea of these mechanisms is to base decisions about whether an interaction between service features has occurred, and how to resolve it, on the actual results of the service operation rather than on some preconceived rules about how services might be expected to interact. This roughly corresponds to the 2-pass feature

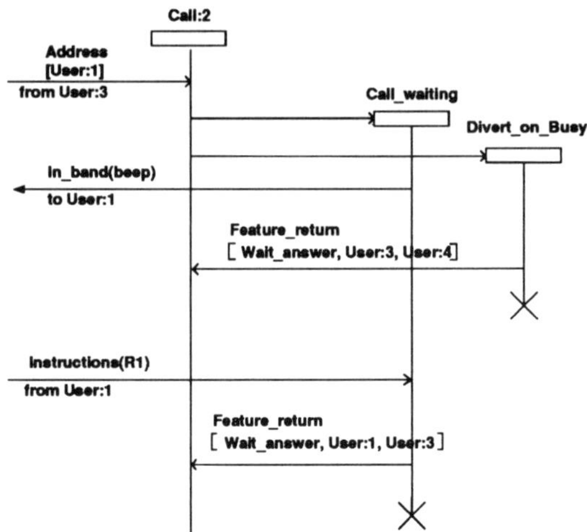

Figure 4, Interaction between Call waiting and Divert on BusyDivert on busy

management described in [4].

3.2.1 Control segment concept

Using the IN concept of "points in call" (PICs) the operation of a call and any associated features can be split into control segments. In our model there is a direct correspondence between segments of the basic call and IN PICs. Also, features are composed of a number of control segments which are managed by service logic. Features may often have only one control segment, but for example in the case of a persistent feature there may be many control segments, one for each time the feature has control of the call. This is illustrated in figure 5.

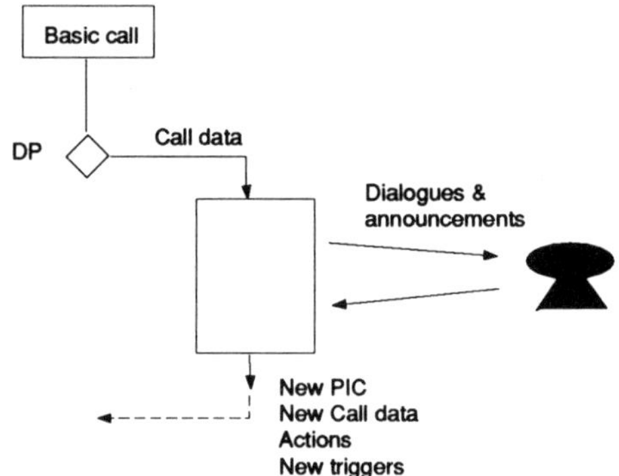

Figure 5, Call control segment

Each feature control segment starts at a basic call Detection Point (DP). If conditions are right for a feature to be triggered, control is passed to the feature along with data about the call. The feature may interact with customers connected to the call, then return control to a basic call control segment, requesting a number of changes to the call, for example :

    * New PIC (e.g. Wait answer)
    * New call data (e.g. connection state)
    * Actions to be taken (e.g. make a new call)
    * New triggering information (e.g. re-invoke this feature when a call is answered)

The feature control segment ends before these requests are acted upon.

3.2.2 Independent feature operation

There is no reason why there cannot exist multiple control segments belonging to different features, but acting on the same call, as shown in figure 6.

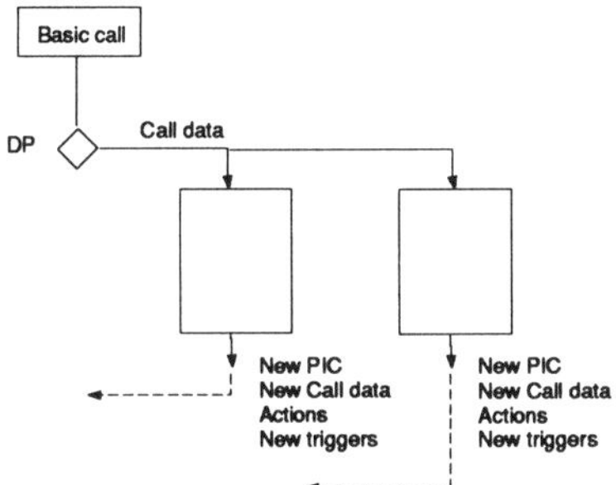

Figure 6, Multiple call control segments

Features which are invoked at a Detection Point can be allowed to operate on their own copies of call and network data independently until there is a need to synchronise at a Point of Return to a basic call control segment. At this point a comparison of the results of the active features on call and network data can be made and a decision made as to whether these results can be accommodated by the network.

3.2.3 Late interaction detection and resolution

The important thing to note about detection and resolution of interactions in this environment is that the process:

    * is based on the actual operation of the feature software for a particular call.
    * uses a common information model of the basic call

This means that it should not be necessary to explicitly specify the interaction avoidance operation of every feature (or even between each possible set of features). Instead, generic processes can be developed which operate on the common information model. This is important in retaining the key commercial benefits of timely feature development. It is even more important where a network supports third party service providers who may not provide access to details of their services.

Also, the strategy for resolving an interaction can use information about the particular customers involved. This consideration appears to be crucial as, in many cases, FI resolution depends on the customers' *intentions* in using the features in question.

3.2.4 Testing

The Service Plane model was first used to investigate feature interactions which were already well known. This early work validated the modelling approach and provided the necessary confidence that interactions which we were previously unaware of could be found with further work.

The next phase of testing looked at detecting interactions involving a fixed network Universal Personal Telecommunications (UPT) service. The model feature set was extended to include this service and a period of testing aimed to uncover any interactions. Here we consider UPT to be a service allowing a customer to receive calls at any fixed line, make Chargecard calls from any line, and have access to a number of supplementary services including call diversion. The UPT service also includes message taking for uncompleted calls. Messaging services allow a caller to leave a voice message which can later be retrieved by the message box owner.

Testing of the UPT trial implementation, based on the model results, identified 7 concrete interworking problems. These involved UPT customers using fixed lines with other services active and are detailed in [7].

One example, illustrated in figure 7, concerns a UPT customer registering his current destination as at a line which has divert immediate active. Here the UPT customer will never receive any incoming calls as they always follow the line based diversion onto another line.

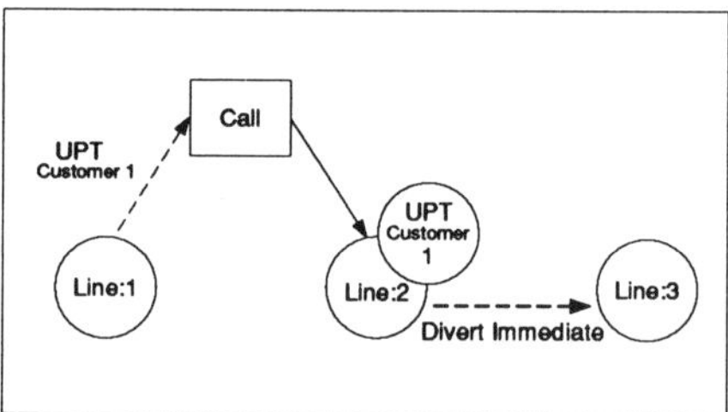

Figure 7, UPT incoming call and fixed line diversion

Another example, illustrated in figure 8, concerns uncompleted UPT calls to lines with fixed network based messaging services. Variations in the programmable delays before each messaging service answers the call can result in messages being left in the wrong message box.

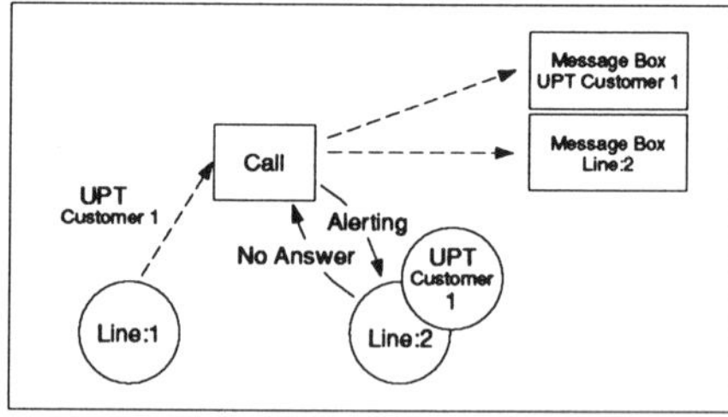

Figure 8, UPT messaging and fixed line messaging

These results concerned problems with the trial implementation which had not been discovered by the usual service development process. The results are being assessed to determine what changes should be made in the service implementation for commercial launch.

3.2.5 Implementation

The implementation of the FI resolution strategy discussed earlier would of course depend on the detailed network architecture. But some general observations are possible based on the IN standards conceptual model. Some knowledge of the IN model is assumed here. Figure 9 illustrates how control segment synchronisation could map onto the conceptual model.

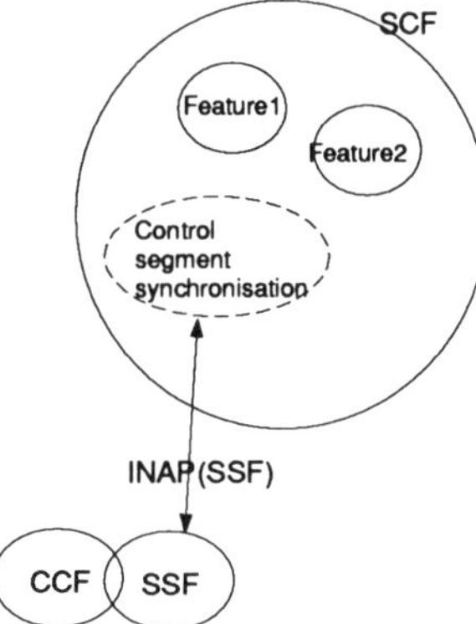

Figure 9, FI method mapped onto IN standards model

Concurrent feature operation for a single call is not ruled out by the standards model. The management of multiple control segments associated with a given feature would clearly reside in the Service Control Function (SCF). Synchronisation of the control segments of different features would have to occur somewhere between the SCF and the Service Switching Function (SSF), which is assumed to be collocated with embedded call control. SCF and SSF can be separated physically by the Intelligent Network Application Protocol (INAP), which in its current form may not support the information flows required for FI resolution. Therefore in the near term feature control segment synchronisation may be more readily supported as an SCF function rather than an SSF function.

4. Results

The results of our work so far have been encouraging. The two aims have been to develop techniques for observing how features will interact within the network and to generate service integration tests for real network implementations.

The process of developing and experimenting with the service models has highlighted that problems can be detected both by static and dynamic observation. The process of model development identified poorly defined parts of the specification where the resultant ambiguity led to feature interaction possibilities. Dynamic detection of interactions was achieved by performing model simulation. In general this approach was able to detect interactions which were difficult to predict prior to simulation.

The highly interactive and visual nature of the animation tool support has proved extremely useful for detecting feature interactions by experimentation with the models.

Where interaction problems have been detected, the SDL model environment has proved a useful vehicle for prototyping any proposed resolution mechanisms.

We have been able to obtain some confidence in our approach and models by 'detecting' known interaction problems as well as some which were previously unknown. We have developed tests for Centrex and UPT service packages that have resulted in problems being uncovered in the delivered implementations. With the steadily increasing complexity of telecommunications service offerings, our modelling approach offers much needed practical support to the necessary service integration activities.

References

[1]   E. Cameron, H Velthuijsen, Feature Interactions in Telecommunications Systems. IEEE Communications Magazine, August 1993, pp18-23

[2]   E. Cameron et al, A Feature Interaction Benchmark for IN and beyond. IEEE Communications Magazine, March 1993, pp64-67

[3]   P Combes, S Pickin, Formalisation of a user view of network and services for feature interaction detection. Feature interactions in telecommunications systems, 1994

[4]   M Cain, Managing run-time feature interactions between call processing features. IEEE communications, February 1992

[5]   M Crowther, Modelling and validation of telephony network signalling. 6th SDL forum, 1993

[6]   L Boollier, B Kelly, et al, Evaluation of some test generation tools on a real protocol exampl. 7th Int. workshop on protocol test systems

[7]   B Kelly, M Crowther, J King, Feature interaction detection using SDL models. IEEE Globecom 1994

A Platform for Modelling Feature Interaction Detection and Resolution Techniques.

Dave Marples*
GPT Ltd, Technology Drive, Beeston, Nottingham. NG9 1LA
DMARPLES@COMMS.EEE.STRATH.AC.UK

Simon Tsang, Evan H. Magill, D. Geoffrey Smith**
*University of Strathclyde, Dept. of Electronic & Electrical Engineering, Royal
College Building, 204 George Street, Glasgow. G1 1XW*
STSANG@COMMS.EEE.STRATH.AC.UK

Abstract

A platform to support the development of experimental call models is presented. The platform, known as DESK, has been created specifically to aid the testing of novel Feature Interaction Detection and Resolution techniques.

The paper presents an overview of DESK before going on to consider the detail of its implementation and the communication between its component parts. It then goes on to consider the interface between DESK and the call model. Two typical call models that have been implemented using DESK are discussed; a simple Monolithic Call Model and one based on the ITU-T Intelligent Network recommendations.

The paper concludes that DESK speeds and simplifies the development of experimental call models and is particularly useful for presenting results in a form which is easy to visualise and relate back to the real world. References are included to papers which contain results produced using the platform.

1. Introduction

This document describes an experimental testbed which has been developed to support research into the Feature Interaction problem.

The testbed, known as DESK, is used to support the development of experimental call models. It has been used in a number of experiments both at the University of Strathclyde and by GPT Ltd and the results of some of these have previously been published.[1,2]

Developing a practical call processing platform in order to test feature interaction detection and resolution techniques is an extensive undertaking. By the use of DESK this task has been simplified allowing the researcher to concentrate on the task in hand rather than the infrastructure to support it. The platform has been found to be particularly suitable for use in Feature Manager implementations such as those in the literature.[3,4,5]

The paper begins with an overview of the platform before continuing with the consideration of typical call models that have been implemented on it, including an IN Model developed by the University of Strathclyde. It then goes on to discuss observations that have been made during the development of the system before presenting conclusions on the effectiveness of such a tool in the investigation of the Feature Interaction problem.

* Jointly funded by the Royal Commission for the Exhibition of 1851 and GPT Ltd.
** Jointly funded by the UK Engineering and Physical Sciences Research Council and the University of Strathclyde.

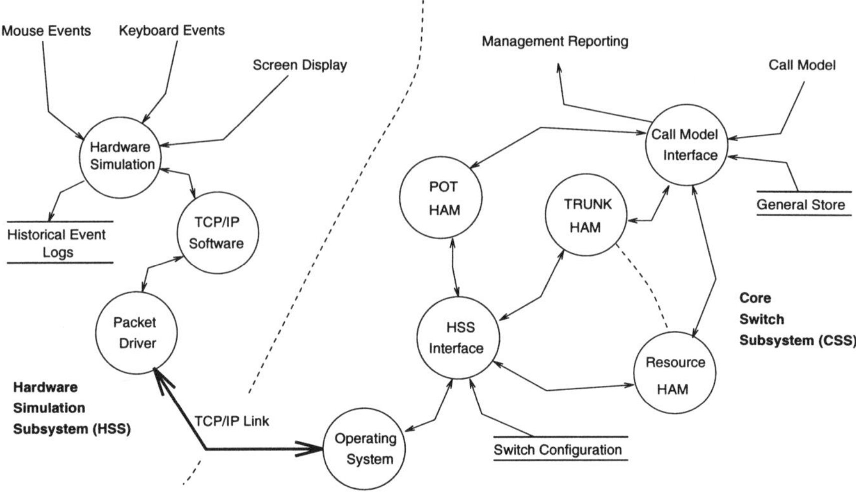

Figure 1 : Structure of the HSS and CSS

2. The Platform

DESK consists of two separate subsystems. The first of these emulates the physical hardware devices of a switching system; the terminals, tone generators, cross connect matrix etc. It is known as the Hardware Simulation Subsystem (HSS). The second subsystem implements the interface to the HSS and forwards switching events to call model implementations. This is the Core Switch Subsystem (CSS). The HSS and CSS are implemented on separate hardware platforms. Both of these subsystems and the communication between them will be considered before going on to discuss the communication between DESK overall and the experimental Call Models that have been developed using it. The split between the HSS and the CSS is shown in Figure 1.

2.1 Hardware Simulation Subsystem

The Hardware Simulation Subsystem (HSS) emulates real hardware devices as closely as possible. It runs on a IBM™ PC MSDOS™ platform and communicates with the CSS via a TCP/IP communication link over Ethernet, using a Packet Driver[6] and Waterloo TCP[7] libraries. The 'hardware' is displayed graphically in a window on the computer and is manipulated by means of a mouse. A typical example is shown in Figure 2.

2.1.1 Implementation Description

Realism considerations received major emphasis during the development of the HSS. Its design was based on the implementation of real switching systems and a design target was that the software simulation should be capable of being replaced with real hardware with no changes to the call model implementations. No convenient shortcuts were taken - for example the Cross Connect was implemented completely independently of the individual devices to which it connected. It would have been more straightforward to connect devices directly to other devices and imply the actions of the cross connect, but at the expense of significant loss of realism.

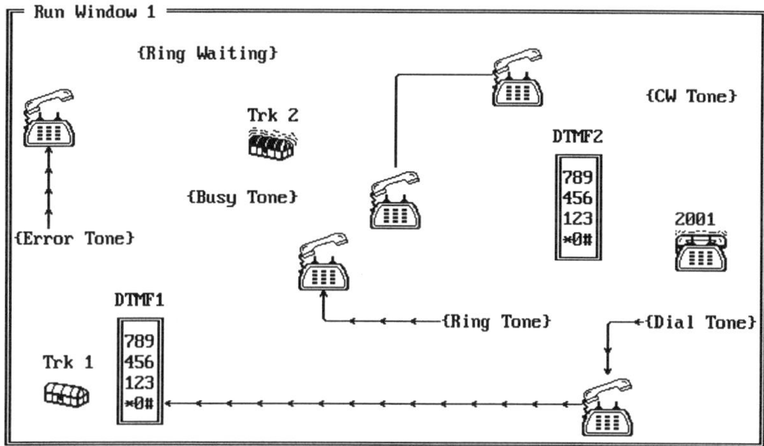

Figure 2 : Example of the HSS in use

There are a number of benefits which have been gained by simulating the hardware; it can be 'created' by means of requests from the CSS to create particular configurations and run conditions. It is also relatively simple to record events from a particular run to be replayed later for analysis purposes.

The exact means by which manipulations occur is dependent upon the device concerned. Changes in their states are reflected by means of differing representations on screen. Examples are shown in Table 1.

2.1.2 Communication

Communication between the HSS and the CSS is via a set of messages which closely match the messages sent between control software and real switch hardware. Additional messages are provided for the creation and destruction of resources.

Table 1 : Elements of the hardware simulation

Plain Ordinary Telephones (POTs) Can go on and off hook and can generate a recall signal. They can also ring when instructed to do so by the CSS.	
Tone Resources. Tone resources are used for communicating states and information to users of POTs. No user manipulation of tone resources can be performed and they only have a single state.	{Busy Tone} {Dial Tone}
Trunks. Trunks are used for external communication. They can be taken into and out of a ringing state and can generate line reversal. The CSS can seize and release them.	
Cross Connect. Connections are uni-directional, they are made from a device to another device and are shown as such on screen with an arrow indicating the direction of information flow. Bi-directional communication (for example a simple POT to POT call in progress) is shown by a communication link with no arrows.	

These messages are divided into classes;

- **Downlink Control Messages:** These are the messages from the CSS to the HSS and are used for modifying the state of the specified resource. Typical messages in this class include RING_POT, SEIZE_TRUNK and cross connect manipulations. There are no commands to enquire on the state of a particular resource. The system is completely event driven.

- **Uplink Control Messages:** These are the messages from the HSS to the CSS indicating that there has been a change in the state of a resource. Typical messages in this class are TRUNK_RINGING and POT_HOOKSTATE.

- **Management Messages:** These are the messages used for the creation and destruction of resources. When a resource is created the appropriate number of cross connect entries are also created and assigned to it. It is then displayed in the run window. When the resource is deleted it is removed from the run window and the cross connect entries are invalidated.

2.2 The Core Switch Subsystem

The Core Switch Subsystem (CSS) software is responsible for the control and management of the interface to the hardware and communication with the Call Model implementation. It initialises the hardware and populates the appropriate resources so that the call model does not need to be aware of the emulated environment in which it is running.

The CSS also provides some facilities which are used by call models. Each of these facilities is described briefly in turn and an example of the communication between them in response to one incoming message is then presented.

2.2.1 Hardware Model Communication

When the CSS first starts, it attempts to make a connection to the HSS. Once this connection has been established a file containing the configuration of the hardware for

```
SECTION SYSTEM

Name "This is the Demo Switch"

SECTION OBJECTS
SR:Dial~Tone
SR:Busy~Tone
SR:Error~Tone
SR:Ring~Tone

POT_SOUNDED : 2001      // This POT has sound support.
POT : 2002
POT : 2003
POT : 2004
POT : 2005
POT : 2006

// These are the trunks in the system
TRUNK : Trk~1 0001
TRUNK : Trk~2 0002

DTMF : DTMF1
DTMF : DTMF2
```

Figure 3 : Hardware Configuration File

this particular run is parsed. The appropriate hardware is configured in the HSS before active operation commences. A typical hardware configuration file is shown in Figure 3. All of the types of hardware that are available in the HSS are configured via this file.

Hardware Abstraction Models (HAMs) are created within the CSS for each piece of 'hardware' that is created in the HSS.

Once all of the hardware resources have been configured, active operation of the system can commence. Messages from the hardware are passed to the HAMs for processing and messages from them are passed back to the hardware.

2.2.2 The Hardware Abstraction Models

The HAMs are a collection of state machines which are abstractions of the current state of each of the hardware devices in the system. They provide the minimum level of insulation from the HSS and they impose restrictions which are defined by the physical realisation of the system.

Figure 4 shows a typical HAM, the Plain Ordinary Telephone (POT) HAM. It has five states and the transitions between them define the physical restrictions of a POT; it is not possible for a POT to enter into a ringing state when it is not on hook, nor can there be more than one connection to it via the switching matrix. It does not prevent any legal transitions from occurring, even if they are normally somewhat inappropriate. For example, it can be seen that it is possible for a POT to be in the Connected From state even when it is On Hook. Normally this would not be a sensible condition but it may be applicable for implementing single party held features, such as emergency call (999/911/112) and some test operations. The CSS will generate an exception indication for any attempt to make a prohibited HAM transition.

Each type of equipment has its own Hardware Abstraction Model (HAM) with its own set of restrictions, capabilities and supported operations. A conference bridge HAM would not support an alerting operation but would permit multiple connections to it, for example.

2.2.3 Call Model Interfaces

Action messages from each of the HAMs are passed to a Call Model Interface. This exports a socket to which call models can connect. The Call Model Interface is responsible for the correct operation of the protocol across this link. Details of the communication protocol are discussed in 2.3.

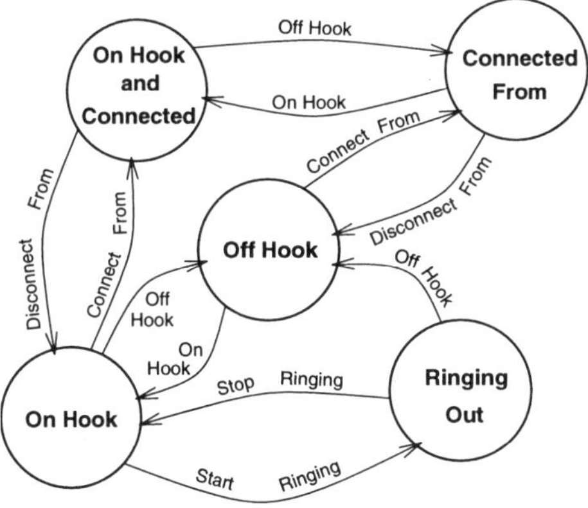

Figure 4 : A Typical HAM

2.2.4 Additional Core Switch Capabilities

Database

The Core Switch contains a database facility to act as a central data repository. It contains information such as population records and management status. Call models can interrogate this database to find out the hardware configuration of the system and they can also add records into it. Some entries are standardised, such as the equipment population records. This allows call models to access this information and Call Model Library routines support this interrogation. The database can also hold records on behalf of other objects within the system. Persistent state information, such as provisioning and charging status, can all be stored. The database is primarily provided to allow information which is of relevance to the physical switching fabric to be co-located with it but it is also used in a more general fashion by call models.

Resource Control

The Core Switch can also be responsible for the control and allocation of resources. Typical of the types of things that are controlled by resource control are DTMF Register, Conference and Protocol Conversion circuits. Allocation of these can be performed on a dynamic basis.

Other Capabilities

There exists within the core switch the ability to add new functionality to support particular requirements or call models. For example, changes were made to the call model interface software in the implementation of the monolithic call model to support a feature interaction manager and feature interaction resolution tables.

The message set for DESK to call model communication has been designed to be extensible orthogonally to existing functionality and additional capabilities can easily be provided within the structure of the existing system.

2.2.5 Message Transfer in Response to Typical Event

It is informative to consider the flow of messages around the system in response to a typical incoming event. The event under consideration is the mouse click on an On-Hook POT, resulting in it going off hook. The information flows are shown in the form of a message sequence diagram in Figure 5.

The mouse click is interpreted by the HSS based on the location on the screen (to identify the resource under consideration) and the current state of that resource. In the example above the requested resource is POT x and the event is the Off-Hook signal. This results in a message being passed to the call model via the HSS Interface, POT HAM and Call Model interface. The Call Model determines what responses are appropriate to the event and passes these back via the same route. The actions in response to the event are then complete and the system returns to an idle condition, waiting for the next event to arrive.

2.3 Interface to Call Models

As has already been mentioned in 2.2.3 the interface between the Low Level Switch control software and the call models implemented on top of it is via a socket based interface. This is exported by DESK on socket 2001 and call models connect to it using standard BSD socket connect semantics.[8]

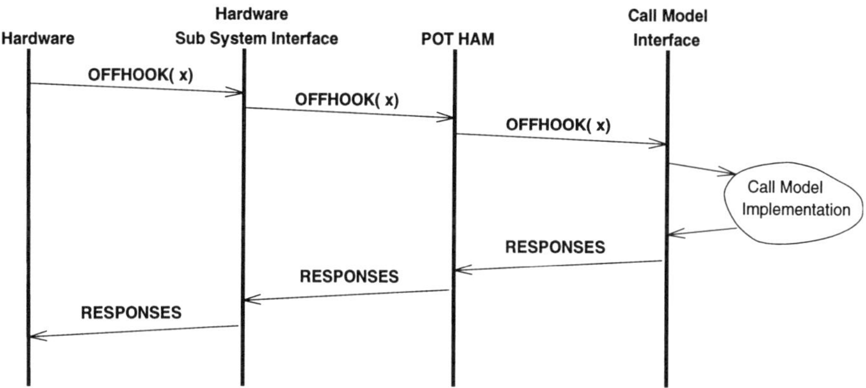

Figure 5 : Events in response to mouse click on an On-Hook POT

There are several classes of messages that are passed across this interface. All are encapsulated in a standard packet format across a TCP link as shown in Figure 6. The messages are subdivided into eight classes;

Uplink (Core Switch to Call Model)

- **Server_Terminating:** Indicates to the call model that the low level switch is about to cease operation. Allows a clean termination without error conditions.

- **HAM_Event:** Event delivery from the Hardware Abstraction Models. Encoded in the fields following this header will be the HAM concerned, the logical equipment number, the event itself and any additional information required.

- **Database_Response:** The response to a database operation. This could be the result of a query or confirmation of a command request.

- **Management_Response:** Response to a management control request. Even though the Call Model is able to treat the Core Switch as being a 'real' switch it is sometimes useful to send control information across the link.

Downlink (Call Model to Core Switch)

- **Client_Terminating:** Indicates to the Core Switch that the Call Model is terminating. Again, this allows a clean termination without error conditions.

- **HAM_Action_Request:** HAM, and hence physical equipment, state change request.

| Length | Class | Item | Message |

Figure 6 : TCP Packet Message Structure

- **Database_Request:** Request to perform a database operation, such as enquiry, update or deletion.

- **Management_Request:** To provide a management control information interface.

These message sets allow complete communication between the Core Switch and the Call Model that is implemented on top of it. There is sufficient flexibility in the system to allow for a wide variety of implementations of Call Models and other experimental constructs.

3. Call Model Implementation on top of DESK

Two different call models have been implemented on DESK at the present time. Both of these have been used in the investigation of the Feature Interaction problem and results based on their use have been published.[1,2]

3.1 Simple Monolithic Call Model

One of the active areas for investigation in the field of Feature Interaction has been in the incorporation of Legacy Systems into Feature Interaction resolution schema. Legacy Systems are typified by having poor documentation and complex, fragile code. It is difficult to 'get into' this code to modify it and the only easy point of access into the system is via the interface to the low level switch and hardware devices. A call model has been developed with these restrictions in mind to perform experimentation upon.

Modifications were made to DESK to allow the implementation of a Feature Manager on top of the basic call model interface. Although the Feature Manager could have been implemented completely separately from the call model, exporting a socket for the features to connect to, this was considered to require a degree of duplication of effort which was unnecessary.

The call model and the results of experimentation conducted using it have already been presented.[1]

3.2 Intelligent Network (IN) Call Model

DESK has also been used to implement a call model based on the ITU-T Q.1214[9] Intelligent Network recommendations. The purpose of the testbed was to experiment with different feature combinations and model interactions, and to investigate different techniques to detect and resolve the interactions.

3.2.1 Software Construction

Object Hierarchy

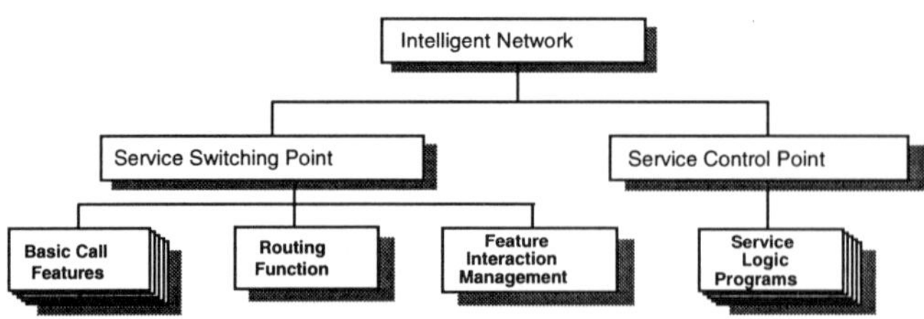

Figure 7 : IN application object hierarchy

The objects in the IN implementation were arranged hierarchically as shown in Figure 7. To create the IN simulation, an Intelligent Network (IN) object is instantiated. This creates Service Switching Point (SSP) and Service Control Point (SCP) objects. Based on information retrieved from the Core Switch database an appropriate number of Basic Call Feature (BCF) objects are created by the SSP and populated with features from a configuration file (the POT-feature profile). Additionally, the SSP creates a special object to provide a platform for Feature Interaction Management (FIM) techniques. Discussion of the available FIM techniques and how they can be implemented using the DESK IN testbed lies outside the scope of this paper.

The POT-feature profile acts as a provisioning table and describes the subscription status for six IN features: Call Forwarding Unconditional (CFU), Call Forwarding on Busy (CFB), Originating Call Screening (OCS), Terminating Call Screening (TCS), Call Waiting (CW) and Automatic Callback (ACB). A typical POT-feature profile is shown in Table 2 below.

A routing function is also created by the SSP object. The routing function provides digit analysis and feature activation/deactivation. A typical routing table is shown in Table 3.

The POT-feature profile is read and the key sequences to activate each subscribed feature are added to the routing table. Note that SLPs are not created until activated by the user, only the trigger points are created. The hardware resources available to the IN simulation are configured by DESK and the IN simulation can be easily scaled up to emulate a larger network by manipulating the configuration files.

Table 2 : IN application POT-feature profile

POT #	BCF #	CFU subscribed	OCS subscribed	TCS subscribed	CW subscribed	ACB subscribed	CFB subscribed
1	1	No	No	Yes	Yes	Yes	No
2	2	No	No	No	No	No	No
3	3	Yes	Yes	Yes	No	No	No
4	4	No	No	No	Yes	No	No

Table 3 : IN application routing table

Directory Number	Action	To Equipment Type	Equipment Number
2001	connect to	POT	1001
2002	connect to	POT	1002
2003	connect to	POT	1003
2004	connect to	POT	1004
*7	activate feature	SCP	Call Waiting
R#7	deactivate feature	SCP	Call Waiting

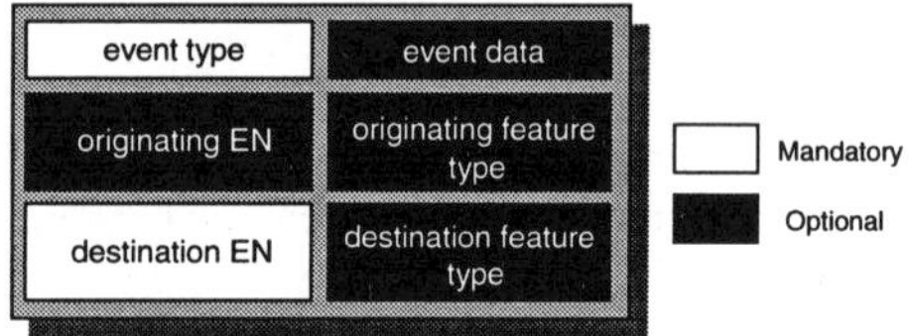

Figure 8 : IN application message type

Integrating the IN Model with the DESK HAMs

State changes in the HAMs are reported to the Intelligent Network (IN) call model by the DESK call model Interface. These events are then passed to the Service Switching Point (SSP). Information about the state of the HAMs is maintained by the SSP. All accesses to the HAMs (such as requests for connection status change) are made through the SSP to them. No other entities can manipulate them directly.

Events in the IN Application

The IN objects intercommunicate using the common message structure shown in Figure 8. A message requires the minimum of an event type and a destination equipment number (EN) before it can be sent. If no destination feature type is specified, it is assumed that the message will be sent to the Basic Call Feature object at the destination address (EN). In addition to this, the originating EN, originating feature type, event data and a destination feature type can also be specified. SLP objects cannot communicate directly with each other within the SCP, they must intercommunicate via the message interface of the SSP.

3.2.2 The Basic Call

A POTS call is handled completely by two Basic Call Feature (BCF) objects. The BCFs are created by the SSP - one for each user. Each BCF acts as a state machine, its behaviour closely following the Basic Call State Model (BCSM) shown in Annex A of Q.1204.[9] In this model, the Basic Call Features can act as either the originating or terminating BCSM (though not both simultaneously) depending on whether the user's

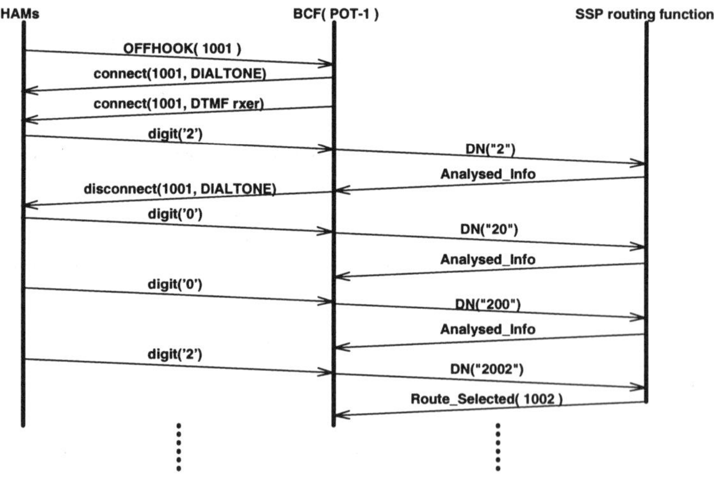

Figure 9 : Initial phase of IN application call set-up - dialling the number

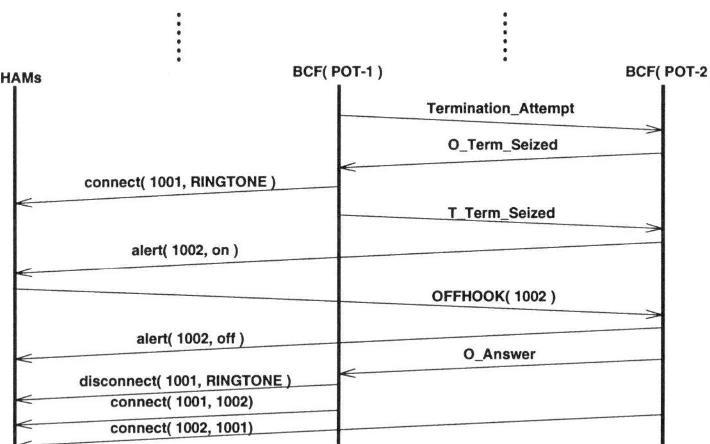

Figure 10 : Completion of Call Setup

POT originated the call. The BCFs are provided with information about the state of the user's POT from the SSP and may send/receive messages to/from other BCFs. BCFs can also manipulate the ringstate of the user's POT and request connection from other network resources. Figures 9 and 10 show a typical call set up scenario between POT-1 and POT-2 with equipment numbers 1001 and 1002 and ring numbers 2001 and 2002 respectively.

Figure 9 shows the initial phase of the call set-up; dialling the number. During this phase, the Basic Call Feature (BCF) interacts with the HAMs and SSP routing function. Note that the routing function performs as a digit translation table during this process.

Figure 10 shows the interactions between BCFs which complete the call setup. The IN testbed in its current form does not model the lower level mechanisms for sending a message from one part of the network to another. It continues from the **Route_Selected** message of Figure 9.

3.2.3 SLP Triggering

Within each Basic Call Feature (BCF), there exists a Detection Point (DP) mechanism which allows SLPs to request and receive certain events. This mechanism is similar to the one described in ITU-T Q.1214, but there are a few differences:

1. ITU-T Q.1214 lists a large number of different "criteria" which must be satisfied before an armed DP will trigger a SLP. This IN call model only allows a Point In Call (PIC) criterion to be specified. The PIC is represented by the BCF state. Criteria relating to digit sequences are specified and handled in the routing function.

2. ITU-T Q.1214 describes two classes of DP: event and trigger DPs. Trigger DPs can only be armed by a management system while an event DP can be armed by SLPs (through an IN service control relationship). The IN application makes no such differentiation, having a single class of DP which may be used by any object in the model.

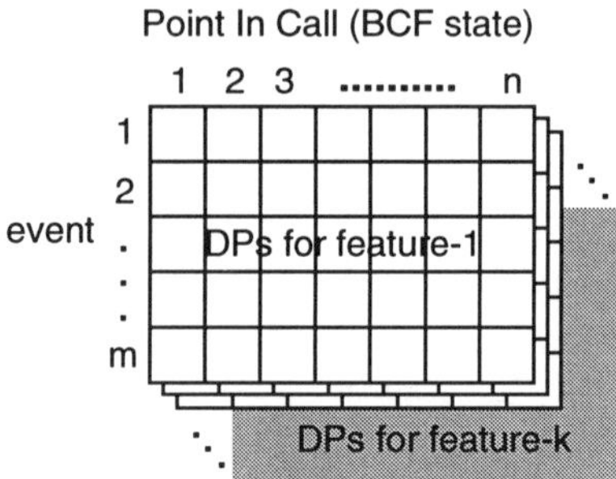

Figure 11: IN application DP data structure

As described in CS-1, each DP can be armed as a Notification and/or a Request DP. The Notification type, when triggered, does not result in suspension of call processing. The Request type suspends the normal call processing function, allowing an SLP to take control of the call. Additionally, each DP can be armed for specific SLPs or for all SLPs. To represent the arming status of the Notification and Request DPs, the IN model uses the data structure shown in Figure 11.

This structure allows DPs to be multiply armed for different SLPs (that is, a DP with the same PIC criteria may be armed for more than one SLP). If a multiply armed DP is triggered, the SLPs receive events in order of their priority. These priorities are currently hard-coded into the application. In this IN model, Notification DPs are processed before Request DPs to ensure that notifications are always provided regardless of the effects of the Request DPs.

When a DP is triggered, it sends a message as described in section 3.2.1.3 *directly* to the SCP object. The SCP object then determines which SLP object is to receive the message and passes it to the appropriate one.

3.2.4 SLPs in the IN Application

The current IN model contains six SLPs: Call Forwarding Unconditional (CFU), Call Forwarding on Busy (CFB), Originating Call Screening (OCS), Terminating Call Screening (TCS), Call Waiting (CW) and Automatic Callback (ACB). The SLPs in the IN model were designed and implemented as state machines but may be implemented in any appropriate fashion. They can arm and disarm event DPs against which they wish to receive events and also manipulate network resources in the same manner as the BCFs (i.e. make requests to the HAMs).

SLP instances are created by the SCP when a user dials the SLP's feature activation key sequence. For example, Figure 12 shows the event flows to activate the call forwarding unconditional feature (activation sequence "*4"). Note that it is the routing function which makes the "trigger_SLP" request to the SCP. This was an arbitrary decision and it may be more appropriate for the BCF to do this.

A problem encountered while implementing the Call Waiting feature was the need to inject a tone into the call without altering the connection status of the subscriber's POT. The current DESK objects do not allow this to be done. The tone injection was eventually implemented by temporarily splitting the POT from the call while connecting it to the Call Waiting tone. A better solution is to add an audio bridge object to the DESK hardware simulation.

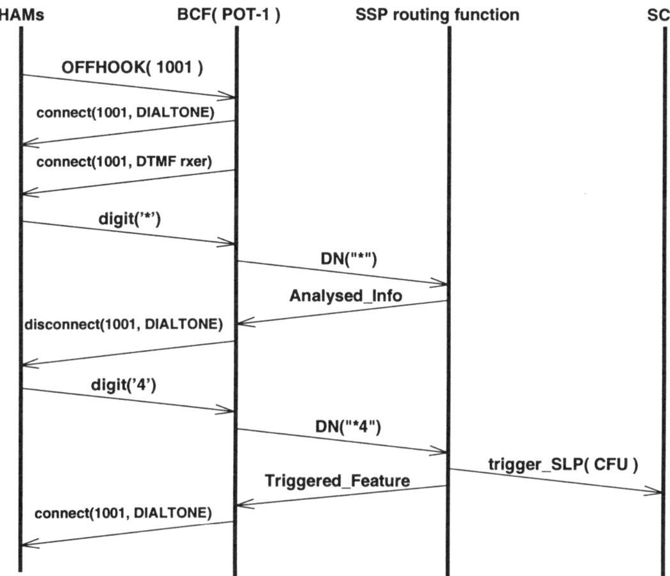

Figure 12 : Activating the Call Forwarding Unconditional (CFU) feature

3.3 Other Potential Implementations

There is no reason why other call models could not be implemented on top of DESK. DESK implements only the physical fabric control and does not constrain the methods or techniques that can be used above. It is intended to be used as a visualisation tool to show the state of network resources in respect of a particular call model scenario and not to constrain the call models that are applicable.

4. Experimental Observations

The single largest practical problem with DESK is its requirement for two hardware platforms on which to operate, one running DOS and another running a UNIX-like OS. This has proven to be a significant practical problem which will probably need to be addressed in the future. It was implemented in this way in order that real hardware could be substituted for the HSS with only minimal changes to the CSS. The demarcation between the HSS and CSS is largely arbitrary and it is unlikely that the move to real hardware will ever be made. At some point the HSS will be moved onto the UNIX system although the HSS/CSS split will probably be retained.

4.1 Processor Requirements

The processor requirements for DESK have proven to be minimal. Since the whole system is event driven and there are no operations to be performed between events there is virtually no processor utilisation while waiting for them to occur. Multiple copies of the CSS have been run on a single workstation with negligible performance impact for other users.

The HSS has been successfully hosted on a number of machines from PC/XT clones upwards. There have been no problems of compatibility to date.

In practice, the single factor that has the largest impact on performance of the testbed is related to the fact that the UNIX-like platforms that the CSS operates on (SunOS 4.x and Linux) do not have any real-time capability, and the response time to an event is largely indeterminate. This is only a problem from a practical system implementation point of view however and has had no impact on the use of DESK as an experimental platform whatsoever.

4.2 Implementation Issues

The use of libraries to aid the implementation of the various call models has meant that most of the communications has been hidden from the point of view of the programmer. Classical distributed computing problems of Endianess and heterogeneous platforms have been completely avoided in this system by means of the use of the libraries. Indeed, we have been able to successfully run the Monolithic Call Model with features distributed across SunOS and Linux machines, and can see no reason why other platforms could not be successfully integrated into the system if required.

With the present HSS/CSS split changes are required to both in order to accommodate new types of hardware, for example conference bridges. It would be useful if descriptions of the hardware were downloaded by the CSS so that the HSS does not need to be recompiled each time new hardware is introduced. This would make the HSS more versatile in responding to changes in the experimental environment.

Most call model implementations assume that the hardware is in a default state when the system starts (i.e. On-hook with no connections between resources). This is an acceptable restriction for an experimental system but would possibly not be suitable for a practical implementation.

5. Conclusions

We have successfully demonstrated the use of a low level switch simulation package which emulates the operation of a real switching platform. It can be used for experimental call model development and it has proven to be versatile and adaptable, easily integrating with existing call model developments. It speeds the development of experimental testbeds and is particularly suited to the implementation of systems which incorporate Feature Manager type functionality.

The fact that the model is based on a real switch is of practical benefit in ensuring that realism considerations are never far away in the minds of the people using it. It is easier to wait five seconds for the results of a run to appear on a display screen than to wait the same length of time for a connect to a dial tone resource after lifting the receiver. The addition of sound, although of little practical benefit to the system, has been extremely useful in re-enforcing the practical nature of it.

We believe that DESK can be used in other environments for the development of other call models on top of a 'real' hardware implementation. At the moment the hardware that is supported is restricted to a POTS environment, but there is no reason why new types of hardware should not be accommodated.

6. Acknowledgements

The help, support and useful comments of the many people at GPT Ltd and the University of Strathclyde during the production of this paper and the ideas which it contains are gratefully acknowledged.

7. References

[1] D.J.Marples, E.H.Magill and D.G.Smith, *An Infrastructure for Feature Interaction Resolution in a Multiple Service Environment, The Application of Transaction Processing Techniques to the Feature Interaction Problem*, Proceedings of TINA '95, February 1995, Vol. 1, pp. 231-242.

[2] S.Tsang and E.H.Magill, *Detecting Interactions in the Intelligent Network*, Proceedings of Second International Workshop on Feature Interactions in Telecommunications Systems, L.G.Bouma and H.Velthuijsen (Eds.), IOS Press, 1994, ISBN 90 5199 165 7, pp. 236-248.

[3] M.Cain, *Managing Run-Time Interactions Between Call-Processing Features, IEEE Communications*, February 1992, pp. 44-50.

[4] L.Schessel, *Administrable Feature Interaction Concept*, ISS'92, October 1992, Vol.2, pp. 122-126.

[5] S.Homayoon and H.Singh, *Methods of Addressing the Interactions of Intelligent Network Services with Embedded Switch Services*, IEEE Communications, December 1988, pp. 42ff.

[6] R. Nelson, Distributor of Packet Drivers, Code Available from ftp://sun.soe.clarkson.edu/pub/ka9q/drivers.exe.

[7] E.Engelke, *Waterloo TCP Programmers Guide*, Release 1.20. Source Code from ftp://dorm.rutgers.edu/pub/msdos/wattcp/wattcp.zip.

[8] W.R.Stevens, *Advanced Programming in the UNIX Environment*, Addison Wesley Professional Computing Series, 1992, ISBN 0-201-56317-7.

[9] ITU-T Q.1214 *General Recommendations on Telephone Switching and Signalling (Intelligent Network), Distributed Functional Plane for Intelligent Network CS-1*, ITU-T, March 1993.

Towards A More Efficient Feature Interaction Analysis - A Statistical Approach

Kristofer KIMBLER

Department of Communication Systems, Lund University
Box 118, 221 00 Lund, Sweden

Abstract: The paper presents an idea of utilizing statistical properties of service usage make the process feature interaction analysis much more efficient. The analysis should focus on combinations of service features which are most probable to occur in practice. The proposed statistical approach, as such, is not any feature interaction detection method. It is a way to simplify the whole interaction detection and resolution process during service creation and to minimize the negative impact of interactions on the service users and subscribers. The paper outlines the statistical approach as well as discusses also the impact of feature interactions on service quality, and the role of feature interaction analysis during service provision.

1. Introduction

The Feature Interaction (FI) problem is commonly regarded as one of the major obstacles to the rapid provision of new services for Intelligent Networks (IN). The operation of newly introduced services and their features alters, often intentionally, the behaviour and properties of the existing ones [7]. The range of problems cause by such interactions varies from an unexpected or undesired service behaviour slightly annoying to the users, to large financial losses, or serious network breakdowns. Inability to detect and resolve such problems might jeopardize the fundamental goals of IN. The problem is not unique to the IN platform and can be encountered in other technical and non-technical domains. In telecommunications, it has been known at least since the first PBXs with a diversity of features were introduced, over 20 years ago.

1.1 Current Research Directions

The FI problem can be addressed in different phases of the service life cycle and on different levels of abstraction. Research in this area focuses on detection, resolution, and avoidance of FIs. Despite a number of significant efforts like [3], no widely approved solutions and no common taxonomy exists in this area, so far. None of the proposed techniques has proved to be able to detect a significant number of previously unknown interactions.

There are tendencies to address the FI problem already in the early stages of service creation, by applying formal modelling of services in SDL, LOTOS, Z [4, 13], or different types of logics [6, 14], combined with automatic or manual verification of their properties. Other efforts concentrate on finding advanced mechanisms, such as the Negotiating Agent Model [8], for resolving interactions during service operation. The proponents of the latter approach stress that on-line resolution provides flexibility and simplifies service creation, which fits the idea of rapid service provision in a multi-provider environment. Also, new concepts of service architectures, such as TINA are being investigated [2]. Their aim is to separate services from the underlying switching capabilities by using, for instance, the concept of Open Distributed Processing (ODP). As claimed, this should increase the orthogonality of different services and features, and thus eliminate, to a great extent, their interactions.

The two latter approaches are the most promising in the long run. However, the need for the detection and resolution of FIs at the early stages of the service life cycle will still exist in the future. Different features, for example, Unlisted Number and Calling Line Presentation, may be introduced to realize contradictory goals, independently of their implementation. Possible combinations of such features and the nature of their interactions have to be analysed during service creation in order to provide appropriate resolution mechanisms or to take appropriate policy decisions. If such policy decisions can lead to restrictions, the service provider should make the user aware of them.

The detractors of FI detection and resolution during service creation contend that this task is a virtually impossible, because from the combinatorial point of view the number of potential feature combinations to be analysed grows at least experientially with the number of features in a network [1]. The reality is even more complicated. Today, we have to deal not only with feature interactions within a single network or system, but with interaction between services and features operating on different networks and different platforms (e.g. IN and GSM). Moreover, in platforms like IN, the same feature can be used as a component of different services (e.g. Speed Dialling is available in Charge Card Calling, Virtual Private Network, and UPT services). Therefore, feature combinations occurring in different service contexts have to be considered as well. Above all, comes the fact that every communication network is the subject of continuous evolution [7], and potential interactions have to be detected and resolved every time a new service or feature is introduced in the network or existing ones are modified.

1.2 A statistical Approach

Growing user expectations and deregulation of the telecom markets cause an increased pressure to rapidly deploy new, technologically advanced telecom services. In order to meet this challenge, service development cycles will have to be radically shorter than today, while the product quality will have to remain very high. In particular, service providers and network operators will have to find time- and cost-efficient methods for detecting and resolving feature interactions, which will enable them to shorten the time-to-market for new services, and to minimize the negative influence of the feature interactions on the quality (availability and reliability) of the services provided.

The complexity of the feature interaction problem mentioned in the previous section combined with growing market pressure on rapid service provision, create a strong need for an approach that will enable both effective and efficient detection and resolution of feature interactions during service creation. In this paper, we would like to address the aspect of efficiency which has not received too much attention, so far.

A statistical usage approach to feature interaction analysis proposed in this paper aims, on one hand, at reducing the number of feature combinations to be analysed during service creation, and on the other, at minimizing the negative impact of interactions on the service users and subscribers. The approach is based on a simple observation that, many feature combinations need not be considered because they never occur in reality or their occurrence probability is extremely low [10, 12]. By focusing on most probable feature combinations, we might reduce the complexity of the feature interaction analysis independently on the detection and resolution methods applied. This is the key idea of the presented statistical approach to the feature interaction analysis.

2. Feature Interactions and Service Provision

2.1 Impact of Feature Interactions on Service Quality

The problems caused by feature interactions, such as unexpected service behaviour, incorrect charging, or inefficient use of network resources, may have a significant impact on the quality of services. If the intensity of such interaction-related problems reaches a certain limit, the services in which the problems occur may be rejected by the service users, subscribes, or even providers.

Each group of actors may have their own reasons for finding a service unreliable. For instance, when 90% users of a Premium Rate service have to wait at least five minutes in a queue, this might quickly discourage them from using this service. Such problem can be caused by an interaction which cause wrong call routing or call distribution. On the other hand, the same Premium Rate service may be received very well by the users (provided the routing problem has been fixed), but at the same time be unacceptable for its provider and subscribes. This might happen when another interaction enables the users to calls the premium rate numbers without being extra charged. This kind of fraud may significantly reduce revenues form the service both for the subscribers and provider.

In general, we may think about the *service quality* as a measure of the user's satisfaction. Of course, feature interaction is just one of the factors influencing this quality. Other factors, such as availability of the network resources, quality of voice connections (often critical in the networks with terminal mobility), or reliability of the service software have also great impact on service quality. The user, is seldom aware of the causes of problems deteriorating service quality. He or she is just concerned that the problems, as such, exist. For instance, the five-minute-waiting in a queue mentioned above, might be equally caused by an interaction, as by an insufficient number of terminating lines where the Premium Rate service is provided. Unfortunately, there are no standardised measures of service quality for the advanced telecom services, especially with respect to the user's satisfaction [12].

2.2 Feature Interaction Analysis as a Part of the Service Creation Process

Deployment of a new service or modification of an existing one might have numerous consequences. Potential advantages and disadvantages of such an operation should be assessed both form the technical and non-technical point of view. This kind of analysis should be an integral element of the service creation process. Feature interaction analysis is a part of it. Service providers, network operators, and other actors have to consider the following correlated factors:

- Market needs and potentialities

- Satisfaction of users and subscribers

- Impact on the quality of other services

- Impact on network performance and safety.

First of all, market analysis should answer the question what is the market potentiality of an introduced service and what are the user's and subscriber's expectations. The predictability of market reactions on new products including telecom services is always limited. For instance, the results of advertising campaigns and other market practices may create social pressure and artificial needs which in a long run may turn to unexpectedly high demand for the services, which happened, for instance, for GSM.

Even if there is a strong market need for a certain service, nobody can guarantee that provision of such a service will be a commercial success. If the overall quality of the service is not satisfactory, the users, subscribers, and even service provider might reject it. This may happen, for instance, when a network cannot cope with a high traffic caused by a rapidly growing number of service users (see also examples in the previous section).

Introduction of a new service can also deteriorate the quality of existing services. One of the potential reason of such deterioration is, of course, feature interaction. Since new services and features may influence existing ones, the users and subscribers may regard the problems caused by the interactions between these services as deterioration of the quality of the existing services. In [12], a method for assessing the impact of feature interaction on service quality is proposed.

Every newly introduced service may also influence network performance and possibly compromise it safety, because it uses the same network resources as existing services. For instance, introduction of a Televoting service may overload periodically a Service Control Point, if it is not protected by an overload control mechanism. Feature interactions may be a cause of potential problems here. The analysis of the service's impact on network performance and safety should be done both for the network in which the service is introduced as well as for other networks interworking with this network.

The question remains, who is responsible for service quality. This question is especially important in the outset of third-party service provision. Today, it seems that every actor involved in service provision should make a relevant part of the risk analysis, however the service quality remain in the responsibility of the one from whom the services are subscribed. The idea of rapid service provision requires efficient approaches and methods for assessing the risk related to the provision of new services.

3. A Statistical Approach to Feature Interaction

3.1 General Considerations

This sections presents a statistical approach to feature interaction problem. The main idea is to utilize the statistical properties of service usage, including probability distribution of the user's behaviour, in order to make off-line detection and resolution of feature interactions much more efficient, as well as to minimize the negative effect of feature interactions on the users and subscribes.

The actual impact of feature interactions on service users, subscribers, and providers is not the same for each interaction case. It depends on the frequency with which a particular interaction occurs in practice and on the scale of damages it may cause. The theory and practice of risk management tells us to concentrate, in such circumstances, on finding and eliminating the interaction cases that are most frequent and most risky. In other words, instead of trying to eliminate all potential interactions, we should rather focus our efforts on limiting the negative effect that interactions have on service users, subscribers, and all the actors involved in service provision. The two examples below will illustrate this point of view.

In the first example, as a result of an interaction, the users of a Freephone service select a wrong service option, because, in some situations, they receive an inappropriate message via Outgoing User Prompter (OUP). If such an interaction takes place in one call out of 10.000, nobody will consider it as a serious error. However, if every second call to the Freephone number is effected by this problem, the service subscriber will not be able to provide the free-of-charge service correctly, and probably will lose a lot of potential customers. This example shows how a minor, but frequent interaction problem may cause a lot of damage.

In the second example, an interaction causes blocking of all calls to a Premium Rate number every time the length of the call queue exceeds a certain limit. Calls to this number can be then rejected for long periods until an action by the service management is taken. Even if such situation is very unlikely to happen, e.g. it occurs once a couple of months, the subscribers will not accept the service in this form, because it may significantly reduce their revenues. Please observe, that if the same blocking problem is automatically removed within seconds, the risk will be marginal as the users will not be able even to notice the problem. This example shows, how a very rare but serious problem may cause a lot of damage.

We may conclude, that a purely statistical usage approach may not be satisfactory for service providers and network operators to assess the risk related to introduction of new services and feature. In general, statistical usage data should be combined with additional factors, such as criticality of service (e.g. use of very limited network resources) or their market potentiality (e.g. possible revenues generated by introduction of the service or feature).

We may use a scale (e.g. from 0 to 10) to quantify these factors for each service and feature. These factors may be then combined with the statistical distributions represented in a usage profile, giving us a total **risk function**. In this way, features with very low occurrence probabilities, but critical impact on the network performance or billing system will receive proportionally more attention during feature interaction analysis then other, non-critical features with similar occurrence probability. In this paper, however, we will concentrate only on statistical aspects of service usage, leaving the risk function as a subject for future research.

3.2 Principles of the Statistical Approach

The idea of the statistical approach to feature interaction origins form usage modelling of services and from Statistical Usage Testing (SUT). Usage modelling of telecom services, though without statistical aspects, has already been successfully applied for detection of interactions between IN services [11]. SUT is a specific type of functional black-box testing in which test data are selected from the input domain according to probability distributions [15]. The concept of SUT comes from the Cleanroom software engineering. A similar method called Operational Profile Testing (OPT) has been developed by AT&T in the field of switching systems [16]. We will use term *statistical testing* while referring to both techniques. The basic idea of statistical testing is to match the actual usage profile during the testing process. Early system exposure to realistic usage brings objectivity and realism to the testing process and helps to achieve significant reliability improvement during testing.

As stated in [12], to enable statistical testing a realistic **usage specification** has to be created. The usage specification consists of the usage model and usage profile. The **usage model** describes the structural aspects of the system usage, i.e., in the case of telecommunication systems, possible sequences of service and feature invocations generated by the user, The **usage profile** quantifies this usage in terms of the probability distribution of these invocations in specific environmental conditions. In other words, a usage specifications describes how the users can employ and combine services and how often certain combinations may occur in practice. The OPT technique proposes a systematic, top-down methodology for creating usage (operational) profiles for telecommunications systems [16].

Statistical testing is a relatively new technique which is just beginning to be used on the industrial scale. Some interesting applications, such as [16], shows that statistical testing can be 20 times more effective in finding execution failures than conventional testing approaches, and the number of user-reported failures can be reduced by a factor of 10. Since the problems caused by FIs are just a subset of all possible system failures, we should expect that also in this area statistical testing will have similar, positive effects.

The real problem with statistical testing is, that it relies very much on the usage specification of a system. Wrongly defined usage profiles might result in an irrelevant reliability assessment. Creation of a relevant and realistic usage specification for newly provided services and features may present a great challenge, requiring extensive market research, prototyping, network traffic measurements, etc. This is especially difficult when new types of services are provided, for which no realistic usage data exist.

A systematic creation of usage profiles for telecom systems requires a thorough market analysis prior and during service creation as well as periodic sampling of traffic in the network. Since the number of users and subscribers of certain services changes in time, so does the usage of these services, the statistical data have to be updated continuously so that it matches the actual state of the service usage.

4. Usage Specification of Telecommunication Services

4.1 Service Usage Modelling Techniques

Statistical approach is based on a usage specification which consists of a usage model and a usage profile. Below, we discuss two usage modelling techniques proposed in [18] and [11] which may be useful in the statistical approach to feature interaction.

The first of the above mentioned techniques applies so-called State Hierarchy (SHY) model which extends the concept of discrete Markov chains. The SHY model has been designed to support the testing of real-time, multi-access telecom systems. The model specifies the usage in a hierarchical manner with services and their features as the lowest levels of this hierarchy. Though the SHY model is designed to support automatic test-case generation, it still preserves the nice analytical properties of Markov chains. In particular, it enables calculation of service occurrence probabilities.

The SHY model assumes that there are different types of system users. A distribution of user types, i.e. a percentage of users in each type, should be proportional the actual traffic they generate. This distribution is modelled on the user type level. Each user type has a number of services available. These are represented on the service level. The actual usage of particular services and features is described on the behaviour and sub-behaviour levels. These levels contain finite state machines which are Markov chains. The SHY model also introduces the notion of time.

The SHY concept is still evolving, and there are a number of problems to be solved. In particular, it has been designed with test case generation in mind, which requires explicit specification of the dependencies between different services and users. Such dependencies are described by means of so-called *links* between different finite state machines on the behaviour and sub-behaviour levels. This means that certain types interactions have to be explicitly specified in the model. This issue must be further investigated if the model is to be used for solving the feature interaction problem.

The second of the techniques does not have the drawbacks mentioned above. It is based on the concept of a Service Usage Model (SUM) developed for the purpose of detection of feature interactions in Intelligent Networks [11]. The SUM enables determination of which combinations (sequences) of service and feature invocations are possible in call scenarios. The model shows many similarities to the SHY model, however, it concentrates purely on service usage modelling corresponding to the behavioural level of SHY. In particular, service features are treated as atomic operations, rather than as separate finite-state automata like in SHY. It is worth noticing, that the SUM-based technique has been successfully applied in EURESCOM Project P230 for detection of interactions between pan-European IN services [10].

A nice property of the Service Usage Model is that each service is described separately by means of Service Usage Graphs (SUGs). A SUG shows explicitly all possible orders of feature invocations within the service it represents. Possible orders of invocations of different services and features in calls are also shown by means of so-called *common states*. This means that no explicit specification of the dynamic relations between different services is necessary. The SUM, as it stands, does not include any statistical profile. The concepts of SHY and SUM have some disadvantages, but a proper combination of their elements may give a usage specification which has both a manageable and simple structure, and good analytical properties. The proposal for such a usage specification is outlined in the next section.

4.2 Creation of Usage Specification

A usage specification cannot be produced by means of ad hoc methods its creation requires a systematic approach. A possible usage specification creation process is shown in figure 1. The creation process starts with the **Use Case Driven Analysis** of service requirements, as described in [9] and [17]. Different call scenarios (use cases) are identified for typical system users (actors). Use cases are described as sequences of events that take place in call scenarios. The use cases are then analysed and merged into a number of Service Usage Models, each is related to one actor (user type). In of some service cases, the Service Usage Graphs (see previous section) may have an identical structures for different actors. It is also possible that SUGs representing the same service have different shapes for different actors, e.g. due to differences in features offered for private and business users.

In parallel to Use Case Driven Analysis, the Statistical Usage Analysis of services is performed. Here a structured approach, like the one proposed in [16] for operational profiles, could be applied. As a result of this analysis, probability distributions of users' behaviour, environmental conditions, and service invocations should be obtained. The statistical analysis is only possible if the service providers and network operators systematically collect information about service usage, e.g. make market analysis and measurements of network traffic.

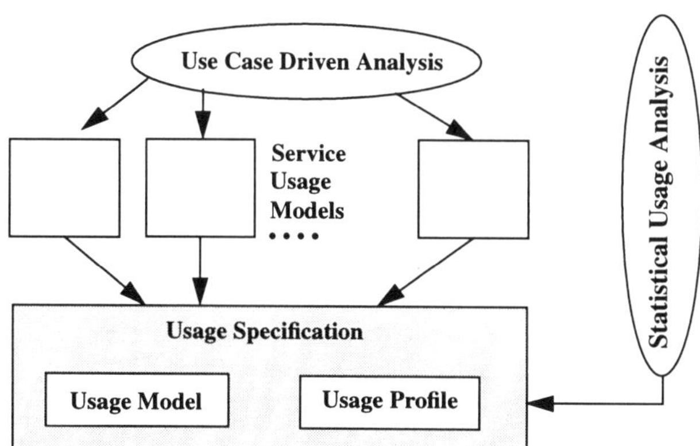

FIGURE 1. Creation of Usage Specification

The SUMs form the basis for creating a hierarchical **Usage Model**. In this sense, the SUMs form the behavioural level of the SHY model. The Usage Model is then complemented with the **Usage Profile** yielded by the Statistical Usage Analysis. The Usage Profile captures statistical usage properties of every service used by each of the user types. The Usage Model and Usage Profile form together the **Usage Specification**.

Note that the same service may have different usage profiles for different types of users. For instance, specific features of the Credit Card Calling (CCC) service may be used with different intensity by private and business users. Hence, the SUG describing CCC may appear in the SUMs related to these two user types, but with different usage profiles.

We should be also aware of the fact that some services have limited target user groups. For instance, while analysing a new service, such as VPN, introduced exclusively for big corporate subscribers we might not consider the part of usage specification related to private users. Of course, the Usage Specification is relevant only for a specific period of time as both structural and statistical aspects of the service usage keep on changing continuously.

5. Potential Applications of Usage Specification

A usage specification of a particular set of telecommunication services can be utilized in different ways:

- It can be used to calculate occurrence probabilities of different combinations of services and features. By focusing on the most probable combinations, and "neglecting" those whose probability is under a certain level or zero, we can dramatically reduce the time and cost of FI detection and resolution during service creation and minimize the negative impact of FIs on the users, and shorten time-to-market for new services.

- It can be employed to generate test cases resembling the real user's behaviour. Such test cases can then be applied for verifying executable models of the system as well as for testing of the system itself. By measuring the mean time between observed FIs (analogous to Mean Time Between Failures), the system reliability can be certified from the FI point of view.

- It can be also applied to calculate invocation probabilities of services and features. This data can provide objective input to analytical and simulation models used for analysing network performance (after introduction of new services). Please notice, that a feature which involves many network resources and very complicated signalling scheme, but is invoked extremely rarely will have only a marginal effect on the traffic and thus on performance of the whole network.

Since most of the existing formal service modelling techniques take the user's point of view [5, 6, 14], we believe that the usage specification outlined in the previous section can be a step towards the creation of more detailed, formal models of services. In this way, consistency and traceability between the usage model and the formal model of services could be assured. Of course, the ultimate goal for the proposed statistical approach is to develop a formal service model that will also cover statistical properties of service usage. In figure 2, a possible role of the usage specification and the formal service model in feature interaction detection and resolution is shown.

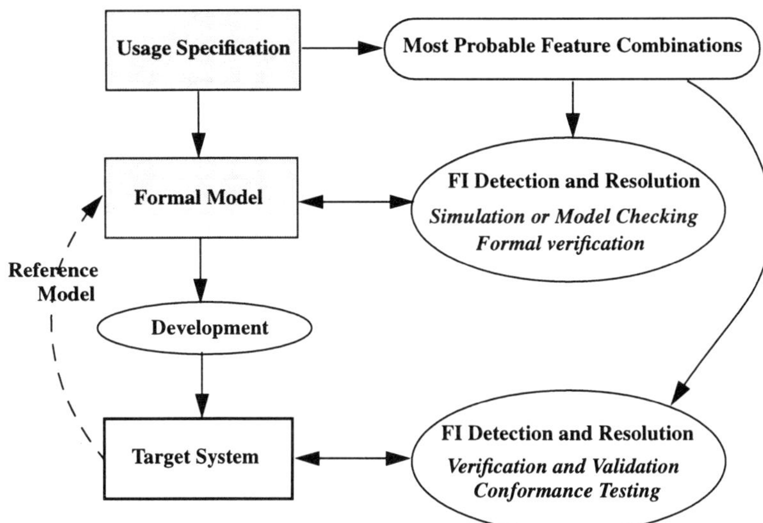

FIGURE 2. Role of Usage Specification in FI detection and resolution.

The statistical approach gives an added value to formal FI detection techniques. By applying the usage specifications to select the most probable feature combinations and call contexts in which their properties have to be verified, the number of combinations to be checked either manually or automatically will be significantly reduced. In consequence, the time necessary for FI analysis will become fairly reasonable, so that it can be performed every time a new service or feature is added to the network. Moreover, by resolving first the FIs which might occur with the highest probability during service operation, the impact of the FI problem on the users will be minimized already at the early stages of service creation. Features with very low occurrence probabilities, but with critical impact on the network performance or billing system, may receive a special treatment (see next section).

Even if we manage to detect and resolve all the most probable interaction cases existing in the formal model, we still have to verify the conformance between the real service software and the service model. In particular, extra interactions that have been introduced during the design and implementation of the service software have to be detected. Unfortunately, exhaustive simulation or model checking are not applicable to the target systems and probably never will be. The only means we can use for this purpose is testing. Testing; but not the exhaustive, time consuming system testing which is commonly used for detecting feature interactions, for instance, in today's switching systems.

No testing technique can assure total coverage of all feature combinations under all possible environmental contexts. This is virtually impossible. The Usage Specifications can be applied to select test cases covering the feature combinations, environmental conditions, and call scenarios with highest occurrence probability, as shown in figure 2. By prioritising such test cases, we would be able to perform the system and conformance testing in a reasonable time, and achieve a very good coverage with respect to the service usage. The formal models of the services could be used as a reference model for automatic verification of the test results, which should speed up the testing process even more. This is however a subject for further research.

In [12] an idea of FI-oriented certification of service reliability is also outlined. The Usage Specification can be employed to automatically (randomly) generate test cases resembling the real user's behaviour, i.e. sequences of the user's actions. Such test cases can be applied for verifying executable models of the system as well as for testing of the system itself. Randomized automatic generation of the test cases from Usage Specification makes the testing process objective, and guarantee that the selected sequences of user's actions will be a representative sample of the user behaviour. By measuring the intervals between observed FIs (analogous to Mean Time Between Failures) and putting this data into a reliability growth model, the system reliability can be certified from the FI point of view.

6. Summary and Conclusions

The idea of a statistical approach to the feature interaction analysis presented in this paper is fresh and still untried. However, the most important elements of this approach have already been practically verified with positive results. To summarize, we can say that the statistical presented approach promotes the following three ideas:

- Statistical usage modelling of telecom services,

- Statistical approach to feature interaction analysis,

- Feature interaction analysis as an element service provision.

The first idea still requires further research and case studies before a usage specification that satisfies all the needs of the statistical approach to feature interaction can be achieved. An ultimate model should capture procedural, non-procedural, and statistical usage properties of the services and their features, and it should also demonstrate a proper balance between the expressive and analytical power, i.e. automatic calculation of service occurrence probabilities, automatic generation of test cases, and automatic detection of feature interactions should be possible.

The second idea, could become one of the most radical solutions for the feature interaction problem. The statistical approach is complementary to other methods aimed at feature interaction detection during service creation. By utilizing statistical properties of the service usage, we are able to dramatically reduce the number of service and feature combinations to be verified and feature interaction cases to be solved. The biggest obstacle here is the lack of proper statistical data concerning the usage of new services. Therefore, to implement this approach in practice, the process of systematic collection of such data has to be started by service providers and network operators as soon as possible. Our job, as researchers, will be to develop a relevant, mathematical models capturing this statistical data, and to provide methods for creation and analysis of these models.

The third idea addresses the problems of service quality management related to provision of new, advanced services. Since feature interactions caused by provision of new services and features, may deteriorate the quality of existing services and the performance of the network, the risk related to this provision must be adequately assessed prior to the service deployment. We believe that feature interaction analysis should become and integral element of risk analysis during service creation.

The final conclusion is, that the proposed statistical approach can be a significant step towards increasing the efficiency of feature interaction analysis, minimizing the time-to-market of new advanced telecommunication services, and improving their quality. If consistently implemented in the service creation process, it can give the service providers a competitive edge in the growing market of advanced telecom services.

References

[1]　T. F. Bowen, C. Chow, F. S. Dworak, N. Griffeth, G. E. Herman, Y. Lin., The Feature Interaction Problem in Telecommunication Systems. In Proceedings of the Seventh International Conference on Software Engineering for Telecommunication Switching Systems, July 1989.

[2]　M. Chapman, P. Farley, R. Minerva, A. Oshisanwo, ROSA - A Service Architecture for TINA. In Proceedings of TINA'93, 1993.

[3]　E. J. Cameron, N. Griffeth, Y. Lin, M. E. Nilson, W. K. Shnure, H.Velthuijsen. A Feature Interaction Benchmark for IN and Beyond. In IEEE Communication Magazine, March 1993.

[4]　P. Combes, M. Michel, B. Bernard. Formal Verification of Telecommunication Service Interactions using SDL Methods and Tools. In SDL'93: Using Objects, Elsevier Publishers, 1993.

[5]　P. Combes, S. Pickin, Formalisation of a User View of Network and Services for Feature Interaction Detection. In Feature Interactions in Telecommunications Systems, IOS Press, 1994.

[6]　A. Gammelgaard, J. E. Kristensen. Interaction Detection, a Logical Approach. In Feature Interactions in Telecommunications Systems, IOS Press, 1994.

[7]　N. Griffeth, Y. Lin, Extending Telecommunications Systems: The Feature Interactions Problem. In IEEE Computer, August 1993.

[8]　N. Griffeth, H. Velthuijsen, The Negotiating Agent Model for Rapid Feature Development. In Proceedings of the Eight International Conference on Software Engineering for Telecommunication Systems and Services, March 1992.

[9]　I. Jacobson et al., Object-Oriented Software Engineering, A Use Case Driven Approach, Addison-Wesley, 1992.

[10]　K. Kimbler, E. Kuisch and J. Muller, Feature Interactions among Pan-European Services. In Feature Interactions in Telecommunications Systems, IOS Press, 1994.

[11]　K. Kimbler, and D. Söbirk, Use Case Driven Analysis of Feature Interactions. In Feature Interactions in Telecommunications Systems, IOS Press, 1994.

[12]　K. Kimbler, and C. Wohlin, A Statistical Approach to Feature Interaction. In Proceedings of TINA'95, 1995.

[13]　A. Lee, Formal Specification - a Key to Service Interaction Analysis. In Proceedings of the Eight International Conference on Software Engineering for Telecommunication Systems and Services, March 1992.

[14]　C. A. Middelburg, A Simple Language for Expressing Properties of Telecommunication Services and Features. In FORTE'94, October 1994.

[15]　H. D. Mills, M. Dyer, R. C. Linger, Cleanroom Software Engineering. IEEE Software, September 1987.

[16]　J. D. Musa, Operational Profiles in Software Reliability Engineering, IEEE Software, March 1993.

[17]　B. Regnell, K. Kimbler, A. Wesslen, Improving the Use Case Driven Approach to Requirements Engineering. Proceedings of 2nd IEEE International Symposium on Requirements Engineering, York, UK, March 1995.

[18]　P. Runeson, C. Wohlin, Usage Modelling: The Basis for Statistical Quality Control. Proceedings 10th Annual Software Reliability Symposium, pp. 77-84, Denver, Colorado, USA, 1992.

Feature Interactions in Telecommunications Systems III
K.E. Cheng and T. Ohta (Eds.)
IOS Press, 1995

Resolving Service Interactions by Service Components

Boris Makarevitch
Helsinki University of Technology, Otakaari 5A, 02150 Espoo, Finland
Email: boris@hut.fi

Abstract: This paper describes an architecture for run-time service interaction management. The architecture is based on the principle of distributed interaction management where interaction resolving mechanisms are not concentrated in a single entity but are distributed between service independent objects (service components). Interactions caused by a specific class of problems are resolved within a component performing service operations from the same problem area. Resolving interactions by the components enhances the possibilities for software reuse while minimising the modifications due to introduction of a new service to the system. A prototype of the model implementing the proposed architecture is presented.

1. Introduction

Intelligent Network (IN) services are usually implemented by sequential programs denoted as service logics. IN service features in most cases can be mapped to parts of the service logics. For example, a service logic for the Virtual Card Calling service can be decomposed to the parts corresponding to the following features: AUTV (Authorisation and Verification), ALT_C (Alternating Charging) and FOC (Follow On Calling). Some complex features such as queuing require a separate program running in parallel with the service logics. Hence, we can identify two distinctive types of interactions: intra-service interactions (i.e. interactions which occur between features residing within the same service logic) and inter-service interactions (i.e. interactions between features residing in different service logics).

Intra-service interactions can be resolved as a result of the service design during composition of a service from multiple service features. It is more difficult to cope with the inter-service interactions since they require cooperation between several concurrent service logics. If the interaction solutions are coded within the service logics, an introduction of a new service to the system may require considerable changes in all service logics to take into account interactions with a new service. Hence, it is more effective not to integrate resolving of inter-service interactions in the service logics but rather to separate resolving mechanisms from the service logics. Placing resolving mechanisms in a dedicated entity in the service execution environment reduces complexity of the problem by separating service actions from service interactions. The service logics programs could be kept relatively simple if they were released from the burden to take interactions with other services into account. Another benefit is that there is no need to change the service logic program every time a new service is deployed in the system.

An entity for interaction resolving mechanisms, named Feature Manager, was proposed in [1] as an intermediate link between the service logics and the Basic Call Process. The Feature Manager's behaviour takes form of a Virtual Machine. This solution can be considered as centralised where all knowledge about interaction resolving is concentrated in one entity. We advocate a distributed interaction management, where several entities resolve interactions intrinsic to their area of concern. The main reasons for that are heterogeneous nature of interactions and a possibility to delegate interaction resolving to service independent objects - service components.

2. Interaction types

Interactions are heterogeneous both by their character and by their domain. Interaction character is determined by when their resolution takes place in the service life cycle. Following this principle we can roughly divide run-time interactions into invocational and executional ones. Resolution of the invocational interactions takes place under service invocation and they are resolved either affirmatively or prohibitively. If the services are mutually exclusive and can not be executed simultaneously invocation of one of them will be rejected. Conversely, the service will be invoked if there are no adverse interactions with other services which forbid its invocation. Obviously, these conditions should be checked during service invocation and resolving of invocational interactions can be located in a unit responsible for invocation of the service logics.

Resolution of the executional interactions takes place during service execution. It is necessary when simultaneous execution of multiple services will bring an undesirable effect. Such resolving can be also considered as service interworking, since with appropriate resolving mechanisms cooperative execution of multiple services will result in a desired effect. When interaction management is separated from the service logics, a service logic does not know about the existence of the other logics and is executed as if it were alone. Then the knowledge about all invoked service logics and how to solve their interactions is located in the Interaction Manager. In the ITU IN Recommendations [3] a unit named Service Logic Interaction Manager (corresponding to the Feature Manager) in [1] is placed in the Service Control Function. However, since the interactions belong to different domains the algorithms and data structures for their resolution are quite heterogeneous. Interaction domains coincide with the major functionalities generic for IN-services. Execution of any IN-service can be decomposed into a set of primitive actions related to the following functionalities:

. user interactions - giving the user some information or collecting data required by the
    service from the user;

. data analysis - processing of data collected from the user as well as data stored in the
    network to enable the service to make necessary decisions;

. connection control - sensing events related to connections and establishing connections on the service demand.

. charging - generation and registration of charging information.

Connection control interactions occur when semantic of the same call event is different for different services. For example, when a Virtual Card Calling (VCC) user with the Follow On Calling feature makes a call to a Premium Rate (PRM) number, an O_MidCall event triggers a new call in the VCC session. This event is unknown for the PRM logic, however the PRM logic should accomplish charging for its call and terminate, since a new call automatically cancels everything related to the old call. Hence, the O_MidCall has the same semantic for the PRM as the O_Disconnect event. According to the interaction management principles the service logics itself should not be aware about events specific to the other services. Therefore resolving of this interaction results in translation of the O_MidCall event to the O_Disconnect for the PRM.

Totally different is the nature of the charging interactions. They occur when service logics executed concurrently have conflicting charging instructions. Resolving of such interactions means resolving of the charging conflicts according to specific rules determined by the network operator or the service provider.

Example: Assume that a VPN (Virtual Private Network) user is restricted by the VPN-subscriber (her boss) to call to PRM numbers. To overcome this restriction the user invokes a VCC service by dialling its access code and call to a PRM number expected to be charged to her VCC account. Obviously, this call should not be charged neither to the VPN account nor to the terminal from which the call was placed. However if the services perform charging as if they were stand alone the following effects may occur:

. VPN has a charging mechanism to register all outstanding (not VPN) calls to a VPN account with a basic call rate.

. VCC has a charging mechanism to register all calls to a VCC account with a basic call rate adding some VCC usage rate per minute.

. PRM has a charging mechanism to furnish a charging record associated with the calling directory number with a premium rate and to register a difference between the premium rate price and the basic rate price as a credit to the PRM subscriber from the network operator.

Apparently the goals of these services are contradicting to each other and without resolving of this interaction in run-time a mess with charging records can be resolved only in court.

The distinctive nature of different interaction groups makes it more rational to integrate resolving mechanisms with the service independent operations and data structures belonging to the same domain, e.g. resolving mechanisms for charging should be located together with generic charging operations.

3. Service Components

The object-oriented paradigm advises to encapsulate all data and operations related to one area of concern into one object. We denote run-time objects containing all data and performing all operations common to a specific domain as service components. The operations located in the service components are often general enough to be included in execution of several services. In other words they can be reused since they are service-independent. Some operations can be specialised for a particular service, however they still include general service-independent procedures of the lower level.

In contrast to the Service Independent Building (SIB) blocks [3] the service components are objects existing in run time. Service components are not just a collection of operations but they have a deterministic control structure in form of a Finite State Machine, where the states decide what operations can be invoked and how they should be executed.

Service components establish a client-service relationship with the service logics. They may have several service logivs as their clients. That enables a component to resolve interactions between the logics which have a relationship with it. The interface between the service logics and service components can be considered as a sort of API (Application Programming Interface) with a language having primitive actions as its lexics.. Consequently, the service logics can be described in terms of primitive actions and primitives defining a control flow of the logics. The relationship between service logics, service components and the other functional entities, as well as their levels of abstractions are shown in figure 1.

As it can be seen from the figure the model has two levels of API. The lower level implements communication between the Service Control Function (SCF) and the other Functional Entities: Service Switching Function (SSF), Service Data Function (SDF), Specialised Resource Function (SRF). This communication is accomplished by exchange of information flows which are directly reflected in the Intelligent Network Application Protocol (INAP) [2].

The upper level provides an interface between high-level service programs (service logics) and the service components. Service logics invoke primitive operations from the service components either in asynchronous or synchronous fashion. Since most of the Capability Set 1 (CS1) services include similar behavioural patterns, the complete set of the operations can be kept rather small. The operations can be classified into four groups according to the domains listed above and consequently located in the four service components: Analyser dealing with data analysis, Charger taking care about charging, Connector controlling connections and scanning asynchronous call events, Interactor performing synchronous communication with the user by means of the specialised resources. Consequently, the Connector can be assigned a task of resolving the call events related interactions, while the Charger is an appropriate object to cope with the charging interactions.

Participation of the service components in interaction resolving requires from them knowledge about multiple service logics instances. Based on this knowledge the service components can resolve interactions in the desired manner. Usually services interact if they are executed within the same call context. We extend the notion of call context to a more comprehensive notion of the service session which is a user initiated and terminated process during which several services can be executed.

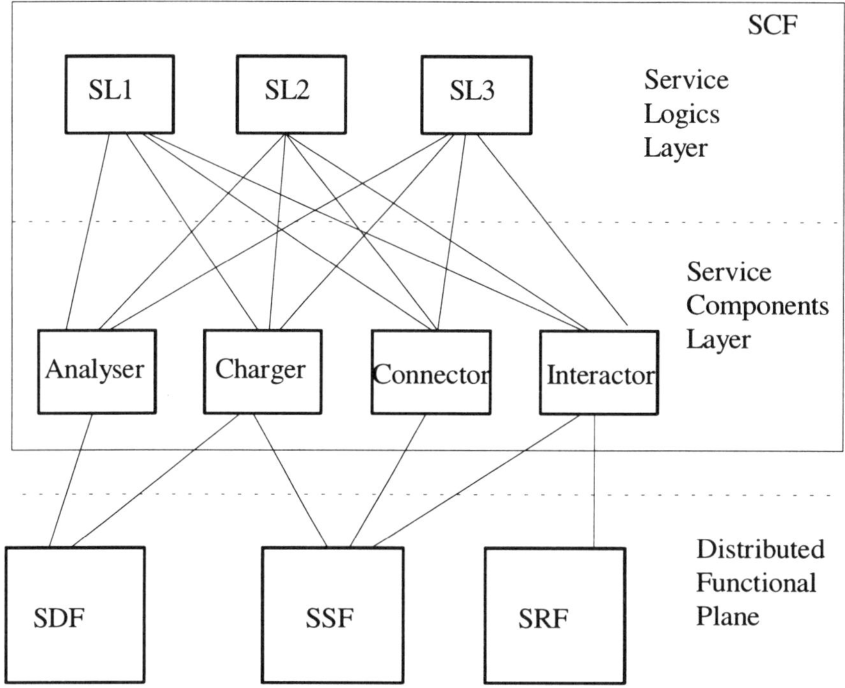

Fig. 1 Service Logics and Service Components Layers

The requirement to have knowledge about multiple services makes a service component unique for the whole service session and equals its life time to the duration of the service session. Thus, service components are created when a service session is initialised and they do not exist after the end of a service session. Since a service component provides an intermediate link between service logics and the other functional entities, it is capable to collect responses from pertinent service logics and to send further a unified interactionless response. Service components are also able to analyse stimulus from the environment and to communicate them to the service logics in a form which the service logics can understand.

4 Description of the Prototype Model

A model of an IN system based on service components was developed within the EURESCOM[1] project "Pan-European Intelligent Network". The work has been continued within the EURESCOM project "Handling Service Interactions in the Service Life Cycle". The main goal of this research was to study interactions and ways of their resolution by simulating IN-services in an executable SDL-model. The model was prototyped in SDL [5] by using the SDT tool from Telelogic initially for three services: Freephone FPH, PRM and VCC. Later the model was enhanced with the VPN service. This enhancement caused

---

[1]EURESCOM - European Institute for Research and Strategic Studies in Telecommunications

relatively few changes to the model as whole and to the interaction resolving mechanisms particularly, thus validating flexibility of the architecture based on the service components.

The block diagram of the SCF is shown below.

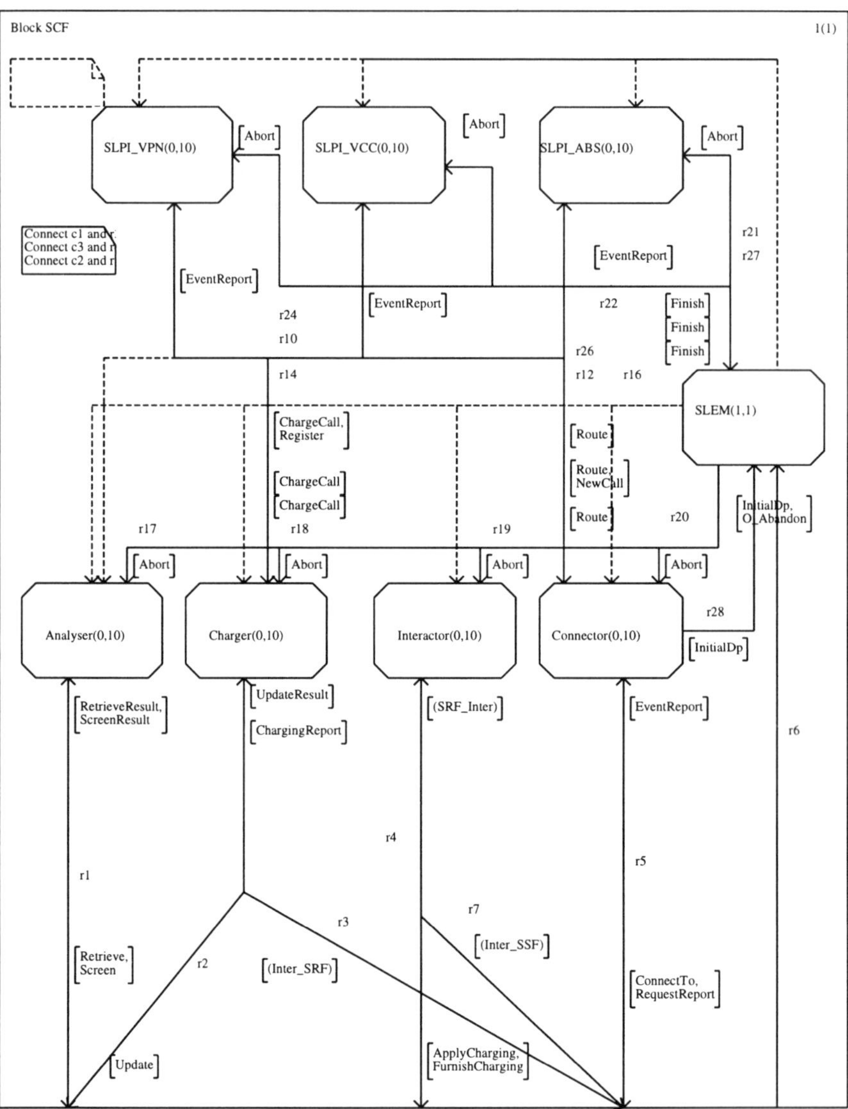

Fig. 2 SCF Block Diagram

The SCF implements four services. Two of them FPH and PRM are executed by the same process SLPI_ABS (Service Logic Program Instance for Alternating Billing Services) as they have identical control flow. VCC is correspondingly executed by SLPI_VCC and VPN by SLPI_VPN.

Operations executed by Analyser and Interactor are implemented in synchronous fashion as Remote Procedure Calls (RPC). They are invoked on request of only one service logic instance and do not have mutual interference.

Operations executed by Connector and Charger (Route, ChargeCall) have usually several service logic instances as clients, during execution of these operations the service interactions between their clients should be resolved. Therefore these operations are implemented in asynchronous fashion by exchange of signals.

SLEM (Service Logics Execution Manager) is a control process responsible for creation and execution of service sessions as well as for invocation of new service logics within the same session. A service session is created upon receiving an InitialDp message from the SSF. Creation of a service session means creation of the service component processes as well as creation of a service logic program instance process for the service requested in the InitialDp message. The necessary data for the service session is stored in the Abstract Data Type (ADT) ServiceSessionsType in the SLEM.

Invocation of new services within the session occurs when Connector detects that the routing number obtained from the user is not a directory number but a service code. Then it sends an InitialDp signal to the SLEM. Since a service session is unique for a specific calling line, the SLEM detects that a session is already created for this line and creates only a requested service logics process. There is one possibility for an invocation reject in our system. It is forbidden to dial off-net VPN access code from VCC. This combination seems to be not practical and it creates additional difficulties in treatment of the O_MidCall event for VPN. If a VCC user dials this code the SLEM will not create an SLPI_VPN process but instead will notify the user by an appropriate announcement via the Interactor.

The session is torn down either upon receiving O_Abandon signal from the SSF or after receiving Finish signal from all the service logics invoked. The Charger component is terminated only if all relevant charging information was updated in the SSF and in the SDF.

Both the call event and the charging interactions are successfully resolved by the service components. Here we demonstrate how the charging interactions are handled by the Charger. The Charger process diagram is shown on the figure 3. The Charger receives from the service logics the signals ChargeCall and Register (to register the accumulative charge for the VCC session). It communicates with the SSF by sending INAP messages ApplyCharging, this message is sent only once for all the service logics involved in the call, and FurnishCharging with final charging information. The Charger receives from the SSF information about the basic call charging record in the signal ChargingReport. Final charging data is sent to be registered in the SDF by signal Update, this signal is embedded in the procedure Update for all the services but VCC.

The rules which govern resolution of charging interactions are encapsulated in the ADT ServiceChargeType. This ADT has two operators *put* and *assign*, which are coded in embedded C. The SDT tool does not support algebraic specifications of the ADTs but gives instead an opportunity to code them directly in C, which is a target language of SDT. The *put* operator is called after reception of an ApplyCharging signal. In addition to storing necessary charging data, *put* sets mutually dependent flags for charging, e.g. if VCC is on, VPN and POTS are off. These flags are later used by the operator *assign* to determine the final charging amount for each service and for the terminal (POTS). It is worth noting that the actual implementation of the rules is not essential for the presented approach. It is more important where an object handling the rules is located.

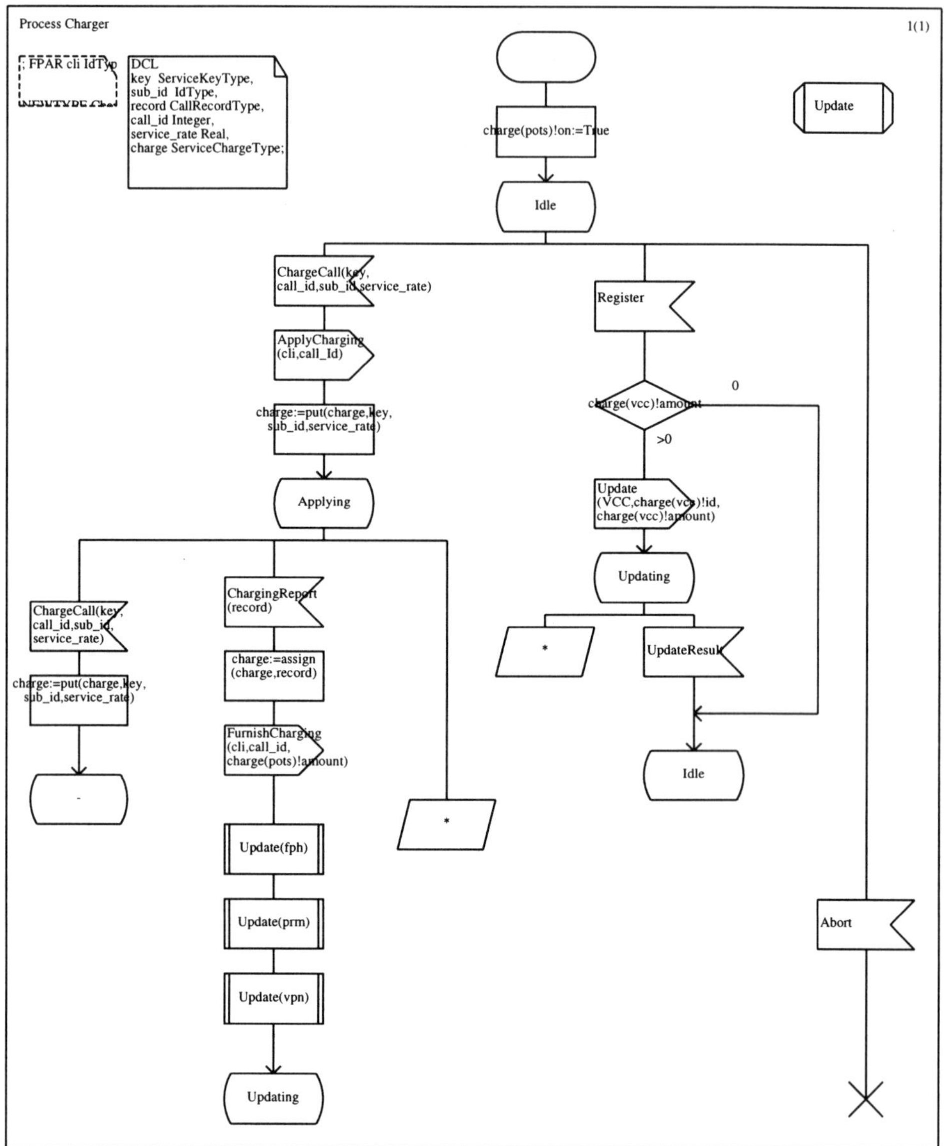

Fig. 3 Charger Process Diagram.

5 Conclusions

Experience from interaction resolution in our model showed that the idea of the service components helps to address the run-time interaction problems in a structured way. Distributed interaction management follows the old proven Roman principle "Divide et Empera". The problem complexity is reduced by the fact that each service component handles interactions specific to its domain. Since interaction resolution is integrated with the low-level service operations a considerable part of the service component specifications can be reused for the purposes of interaction resolution. Introduction of the VPN service to our system did not cause any structural modifications due to the interactions between the VPN and the services already implemented in the system. The rules specific to the VPN interactions were added to the ADT's in which the rules for the other services were stored. Thus, the proposed architecture is very promising for simplifying the task of interaction management. However more experience is needed for further validation and development of the approach. We plan to enrich the model with more services and more service features to gain additional experience.

References:

[1] M.Cain, "Managing Run-Time Interactions Between Call-Processing Features", IEEE Communications, February 1992

[2] ETSI, "Intelligent Network Capability Set (CS1) Core Intelligent Network Application Protocol (INAP), 1993

[3] ITU-TS, Recommendation Q.1213, "Global Functional Plane for IN CS-1", 1992

[4] ITU-TS, Recommendation Q.1214, "IN DFP Architecture for IN CS-1", 1992

[5] ITU-TS, Recommendation Z.100, "Specification and Description Language (SDL)", 1992

Author Index